Critical Power Tools

MW00450837

SUNY series, Studies in Scientific and Technical Communication

James P. Zappen, editor

Critical Power Tools

Technical Communication and Cultural Studies

Edited by

J. Blake Scott
Bernadette Longo
Katherine V. Wills

State University of New York Press

Published by
State University of New York Press, Albany

© 2006 State University of New York

All rights reserved

Printed in the United States of America

No part of this book may be used or reproduced in any manner whatsoever without written permission. No part of this book may be stored in a retrieval system or transmitted in any form or by any means including electronic, electrostatic, magnetic tape, mechanical, photocopying, recording, or otherwise without the prior permission in writing of the publisher.

For information, address State University of New York Press,
194 Washington Avenue, Suite 305, Albany, NY 12210-2384

Production by Michael Haggett
Marketing by Anne M. Valentine

Library of Congress Cataloging in Publication Data

Critical power tools : technical communication and cultural studies / edited by J. Blake Scott, Bernadette Longo, Katherine V. Wills.
p. cm. — (SUNY series, Studies in scientific and technical communication)
Includes bibliographical references and index.
ISBN 0–7914–6775–9 (hardcover : alk. paper) — ISBN 0–7914–6776–7 (pbk. : alk. paper)
1. Business writing. 2. Technical writing. 3. Culture—Study and teaching. I. Scott, J. Blake, 1969– II. Longo, Bernadette, 1949– III. Wills, Katherine V., 1957– IV. Series.

HF5718.3.C85 2006
658.4′53—dc22

2005019152

ISBN–13: 978–0–7914–6775–6 (hardcover : alk. paper)
ISBN–13: 978–0–7914–6776–3 (pbk. : alk. paper)

10 9 8 7 6 5 4 3 2 1

Contents

Foreword
  Alan Nadel ix

Introduction: Why Cultural Studies? Expanding Technical Communication's Critical Toolbox
  J. Blake Scott, Bernadette Longo, and Katherine V. Wills 1

THEORY

Chapter One: The Technical Communicator as Author: Meaning, Power, Authority
  Jennifer Daryl Slack, David James Miller, and Jeffrey Doak 25

Chapter Two: Extreme Usability and Technical Communication
  Bradley Dilger 47

Chapter Three: The Phantom Machine: The Invisible Ideology of Email (A Cultural Critique)
  Myra G. Moses and Steven B. Katz 71

RESEARCH

Chapter Four: An Approach for Applying Cultural Study Theory to Technical Writing Research
Bernadette Longo 111

Chapter Five: The Rhetorical Work of Institutions
  Elizabeth C. Britt 133

Chapter Six: The Study of Writing in the Social Factory: Methodology and Rhetorical Agency
Jeffrey T. Grabill 151

Chapter Seven: Living Documents: Liability versus the Need to Archive, or, Why (Sometimes) History Should Be Expunged
Beverly Sauer 171

PEDAGOGY

Chapter Eight: Writing Workplace Cultures—Technically Speaking
Jim Henry 199

Chapter Nine: Rhetoric as Productive Technology: Cultural Studies in/as Technical Communication Methodology
Michael J. Salvo 219

Chapter Ten: Extending Service-Learning's Critical Reflection and Action: Contributions of Cultural Studies
J. Blake Scott 241

Chapter Eleven: Designing Students: Teaching Technical Writing with Cultural Studies Approaches
Katherine V. Wills 259

Afterword
Diana George 271

Contributors 275

Index 279

List of Illustrations

Chapter 3

Figure 3.1: Schematic Representation of the Evolution of Institutional Frameworks 79

Table 3.1: Habermas's Chart of Work and Interaction (1970, 93) 79

Figure 3.2: Ideological Product Development System 87

Table 3.2: Habermas Translated: Systems of Purposive-Rational (Instrumental and Strategic) Actions in Email 98

Chapter 6

Figure 6.1: Sullivan and Porter Framework for Research Methodology 153

Chapter 7

Figure 7.1: Bernard Challenges Results of Technical Review 181

Figure 7.2: The Cycle of Technical Documentation in Large Regulatory Industries 182

Table 7.1: The CMRR Quantifies Experience 183

Table 7.2: The Process of Transformation 185

Foreword

Alan Nadel

Several years ago, in the Dark Ages of Microsoft Word 4.0 and the Classic Mac, I was putting together a proposal for a conference panel. When the proposal was complete, I ran the "Spellcheck" program that, in the Word 4.0 version, presented one at a time each word that fell outside the Microsoft lexicon. When the word *heteronormative* appeared in my dialogue box, I clicked the "suggest" option, even though I was certain the word was spelled correctly. The box paused, blinked once, and then, under the Apple insignia produced the following message: "'Heteronormative': Can find no alternative."

One might dismiss this as a cute infelicity, but can discourse ever find alternatives to its own lexicon, and isn't language the initial site of power and normativity? Consider that, in American history, whiteness and blackness were not stable biological classifications but shifting sociopolitical categories. Although race designations may be clear examples of cultural meanings, is the meaning of anything completely exempt from the conditions of its production?

If the world is a given, easily predictable and knowable, then technical communication would be unnecessary. Infinite stability would, at least in theory, render everything apparent, self-evident, and hence without the need for the technicality of mediation. Only a world evolving, one in flux and subject to constant vicissitudes, requires technical communication. In such a world, however, technical communication too must be embroiled in the conditions of its own production. To imagine otherwise would be to attribute to the field of technical communication divine authority, that is, to define it as a stable activity that stands outside the chance and synergy of all human endeavor, able to produce without ever being a product, able to mediate reality without ever being subject to mediation. Isn't that exactly the raison d'être of the templates in Microsoft Office—to provide a higher authority for the technology of producing a report or a memo?

In this sense, the hyperpragmatism that has, especially in the twentieth century, provided authority for the discipline of technical communication assumes the self-evident position analogous to theological authority. The notion of the pragmatic, the instructional, or the obvious relies on the repression of any context that might question or undermine the tacit assumptions and privileged narratives that create the category of "evidence" and allow for the objectification of knowledge.

Contemporary technical communication textbooks typically identify four categories of audience as comprising the natural universe of technical communication reception: the executive, the expert, the technician, and the layperson. In order to delimit the audience in this manner, however, the textbook has two options: Either it must indicate that it is incomplete, or it must exclude *itself* from the category of technical communication. The latter choice, of course, is impossible. How could a technical communication textbook not exemplify technical communication, especially if technical communication is so naturally pragmatic? The job of a technical communication textbook is to explain, illustrate, and instruct. It succeeds if it shows one how to do something. If technical communication is a body of knowledge that cannot even be exemplified by a textbook devoted to that knowledge, then we should logically assume that it cannot be exemplified at all.

Since that assumption is completely unnatural, we have to opt for the other alternative: that the book is incomplete, that there is another audience for technical communication that the book—and the field itself—omits, and it is the very audience at whom that quintessential piece of technical communication—the technical communication textbook—is aimed. An omission this large, however, must be more than just a technical error. It is the way of keeping the mechanisms of technical communication invisible by giving the technical communicator a position of authority outside of both the material he or she communicates and apart from the process by which he or she shapes that communication. In this way, technical communication is like Foucauldian power/knowledge system in that it relies on the invisibility of the relationships by which it includes and excludes, by which it orders, measures, and disciplines.

The point I am making is that in any culture and at any historical moment, propositions fall into three categories: those that are viably debatable, those that are not debatable because they are considered ludicrous, and those that are not debatable because they are considered obvious, natural. In this sense, cultural studies is the enterprise of shifting the categories, such that the ludicrous and the obvious move into the realm of the debatable. To put it another way, it is the task set out by Roland Barthes in the preface to *Mythologies* of demonstrating how history and nature are confused at every turn. Cultural studies makes opaque the transparency with which hyperpragmatism functions. It does so in order to reveal the social, ethical, historical, ontological and teleological contexts that permit the illusion of transparency, that is, that substitute faith for knowledge.

It does this in a variety of ways. One is to examine the tenets of institutional practice. Consider, for example, the ways in which the idea of "insurance" requires catastrophe to give it meaning, how it embodies the ideology of protection, although protection would render insurance unnecessary. Insurance therefore substitutes compensation for protection as though the resulting economy comprises a natural equivalency, an illusion relentlessly supported by layers of institutional practice, manifest in—and thus revealed by—every dimension of the insurance industry's rhetoric. That rhetoric, moreover, works to preserve over time the apparent timelessness of its informing institutional claims.

By looking at the historical specificity of any utterance or practice, therefore, a cultural analysis can invoke contexts that redefine the meaning of the utterance or the purpose of the practice. This would be true in technical communication, for instance, of the concept of "usability." Its roots in American pragmatism, Taylorism and Fordism link it to suggestions of ease as well as to the time-motion studies that dehumanize workers, decrease long-term productivity, and foment the conditions that provoke psychotic outbursts of workplace violence, commonly referred to as "going postal." Thus productivity and dysfunction occupy a common industrial matrix.

Cultural analysis can thus reveal groups empowered by an utterance or disempowered through a practice. This is crucial in understanding that legitimation of an institution is the result not of inherent value but of a continuous struggle for the control and distribution of meaning. In this light, the scientific model from which technical communication draws its authority is itself reliant on technical communication to replicate and thus bolster its procedures and the status of those procedures.

Since authority not only requires but creates subjects, cultural studies examines the array of subject positions from which cultural meanings are constructed and construed. As Bernadette Longo states, "a cultural study describes an anti-disciplinary, situated, and personal encounter between a researcher and technical writing practices" (127:69 in original). It can expose the false consciousness that allows people to participate in their own subjugation, and it can help us understand some of the mechanisms that produce false consciousness. Among those mechanisms are cultural institutions and the narratives that they produce. Cultural analysis can help us discover how narratives acquire a cogency that, in turn, produces material effects. A cultural analysis can thus decenter the authority for any assertion to expose what lies invisibly beyond the margins of "nature." In short, it can expose the multifarious ways in which nature is *not* a given.

Technical communication, in other words, is not transparent in relation either to its affect or its referent. The content of textbooks and instruction manuals does not occupy a theological relationship to things it names and describes. Thank God they are all unnatural, that is, all the products of history, and therefore open to historicizing. And thank goodness, we now have a stunning collection of essays that can help remind us of technical communication's positioning in the order of things.

Introduction

Why Cultural Studies?

Expanding Technical Communication's Critical Toolbox

J. Blake Scott, Bernadette Longo, and Katherine V. Wills

How can reviewing technical communication pedagogy, research methods, and theoretical concepts through a cultural studies lens enhance our work and that of our students? The essays in this collection offer a provocative array of answers to this question. Because this question has so rarely been asked, we envision this collection as a sourcebook for the field, surveying a mostly unfamiliar scholarly terrain and providing other scholars the tools with which to continue this expedition. When we talked with colleagues about the idea of this collection, we often encountered the response, "It's long overdue." In most cases, people were responding to technical communication's still largely uncritical, pragmatic orientation, which we will discuss later.

This collection testifies and responds to our field's need for more research and teaching approaches that historicize technical communication's roles in hegemonic power relations—approaches that are openly critical of nonegalitarian, unethical practices and subject positions, that promote values other than conformity, efficiency, and effectiveness, and that account for technical communication's broader cultural conditions, circulation, and effects. In the process, the collection also challenges the disciplinary parameters that have defined and measured our profession and practice in terms of narrow pragmatics and economic success. In addition to being skilled communicators and successful professionals, we and our students must be virtuous citizens who ask critical questions for a sustainable democracy. By furthering this latter ideal, we hope to help heal the rupture in public trust by technical communication mobilized more for profit and greed than the public good (e.g., tobacco and pharmaceutical marketing, finance and accounting reports, energy regulatory standards).

Some readers might be uncomfortable with replacing pragmatic goals with these more normative ones. We are not arguing for total replacement as much as enhancement and redirection, however. The essays in this collection powerfully illustrate how cultural approaches can advance both types of goals; in helping us create more egalitarian mechanisms for producing and assessing texts, for example, cultural studies approaches can better ensure that these texts accommodate their users. That said, we do not necessarily see discomfort as something to be avoided, as it can help us maintain self-reflexivity. Further, if our work were risk-free, it would not likely be transformative.

Readers might also worry about the impulse of some cultural studies work, including work in composition studies, to privilege critical, academic analysis over rhetorical production, an impulse that runs counter to our field's value of civic action. In his essay in the collection's final section, for example, Jim Henry laments cultural studies' "relentless insistence on forming students as critical discursive consumers all the while wholly ignoring their formation as critical discursive producers in any genre other than the academic essay" (215). Several other contributors, including Jeffrey T. Grabill and Michael J. Salvo, echo this healthy concern. Ultimately, however, the editors and contributors to this collection do not see critique/production or academic/civic as either-or issues. In the face of a global corporate culture that, ironically, seems to be narrowing the agency of technical communicators and other workers in the name of flexibility, we need to prepare our students to be both cultural critics and rhetorically effective producers.

Finally, some readers might have qualms about what they see as an imperialist tendency of cultural studies. Carl Herndl and Cynthia Nahrwold voice this concern when they warn against promulgating "theoretical imperialism in which the researcher's theoretical commitment dominates both the scene under study and the social actors in it—a theoretical procrustean bed" (289). Once again, we don't see a conflict of interest in having a theoretical and political commitment and remaining sensitive to the sociohistorical contexts and exigencies of technical communication. As Lawrence Grossberg explains, "Cultural studies is committed to the detour through theory even though it is not theory-driven: it is driven by its own sense of history and politics" (344). Indeed, Grossberg characterizes cultural studies as "radically contextual," though he also insists that the "contexts" it studies are, in part, "defined by the project, by the political question that is at stake" (255). Technical/professional communication scholars Patricia Sullivan and James Porter, in *Opening Spaces*, call for a combination of political commitment and a methodological openness that is shaped partly out of the specific relationships involved in the study. Critical research that aims for ethical action, they argue, must be attentive to the "distinctive nature of writing-as-situated practice" and must define this ethical action not beforehand but "in dialogic concert" with those involved and affected (42–43).

Now that we have explained what, in our view, cultural studies is not (e.g., concerned only with critical consumption, driven only by theory), let us elaborate on how we might define it. Heeding Fredric Jameson's injunction to "always historicize," we begin by historicizing the concept of "culture" and some of the traditions for studying it (*Political* 9). We could start here with classical Marxist theory, the Frankfurt school's neo-Marxist critiques of mass communication and culture, the linguistic structuralisms that have driven so much of U.S. cultural studies, the critical pedagogy of Paulo Freire and other educational theorists, cultural anthropology and sociology, or a number of other places. Indeed, other technical communication scholars, including contributors to this collection, have grounded their explorations in one of more of these traditions.[1] In our framing, we start with Raymond Williams and then move to appropriations of him and others (especially Althusser, Gramsci, and Foucault) by the Birmingham school, as these appropriations serve as the strongest influence on our working definition and approach.

Williams noted that the meaning of "culture" has been shaped by "changes in industry, democracy and class" (xvi) and changes in our "social, economic and political life" (xvii), documenting the shifts in meaning here:

> the recognition of a separate body of moral and intellectual activities, and the offering of a court of human appeal, which comprise the early meanings of the word, are joined, and in themselves changed, by the growing assertion of a whole way of life, not only as a scale of integrity, but as a mode of interpreting all our common experience, and, in this new interpretation, changing it. (*Culture* xviii)

In this passage, Williams forwards a notion of culture that accounts for both modes of experiential living and modes of interpretation through which we make sense of our experiences (Hall, "Cultural Studies: Two" 35). As Stuart Hall elaborates, Williams's culturalist formulation—out of which much of British cultural studies developed—viewed culture

> as *both* the meanings and values which arise amongst distinctive social groups and classes, on the basis of their given historical conditions and relationships, through which they "handle" and respond to the conditions of existence; *and* as the lived traditions and practices through which those "understandings" are expressed and in which they are embodied. ("Cultural Studies: Two" 39)

Williams's impulse to study the relations between material practices/production and meaning-making practices/symbolic production, also discussed in *Keywords*, was taken up by Hall and other critics of the Birmingham school

(90). Other impulses of the culturalist strand of cultural studies (alluded to in the passages from Williams and Hall above) include focusing on concrete, historically contingent practices and allowing for conscious struggle and action in response to "confrontation between opposed ways of life" (Hall, "Cultural Studies: Two" 37, 47).

Cultural studies' (and, more specifically, the Birmingham school's) emphasis on power-laden struggles over competing "ways of life," meaning-making, and knowledge is also endebted to the work of Gramsci and Foucault. Gramsci theorized hegemony as ongoing, shifting power struggles through which certain social groups contingently dominate others. This domination is not just imposed, but won also on the ideological front through "intellectual and moral leadership" that enlists the consent and even participation of those who are subordinate (see *Prison* 57). Gramsci's notion of hegemony enabled Birmingham school critics such as Hall to account for class and other social struggles without resorting to totalizing, universalizing versions of Marxism.

As Longo illustrates in *Spurious Coin*, Foucault's archeological method and theory of knowledge/power provide a basis for assessing how "legitimated knowledges articulated in discourse embody historical [and institutional] struggles for their legitimation and conquest" (16). In addition, Foucault's poststructuralist notion of power as productive rather than repressive has enabled cultural critics to reconfigure subjects as the effects of power rather than just the objects of it. Further, cultural critics have drawn on Foucault's more specific theorizing of disciplinary power to critique the ways "institutional and interpersonal microprocesses" elicit knowledge about people to observe, classify, manage, and shape them as individuals and members of populations (Scott, *Risky* 7). Foucault's later genealogical method built on his archeology by examining both discursive and extradiscursive operations of knowledge/power and by calling for the reactivation of subjugated knowledges. Thus, Foucault contributes to cultural studies' dual emphasis on discourse and materiality as well as its impulse to intervene in hegemonic practices.

Hall points to another important contribution of Gramsci that is also echoed by Foucault. In "Cultural Studies and Its Theoretical Legacies," Hall credits Gramsci's notion and example of the "organic intellectual" with profoundly shaping the work of the Birmingham school. As Hall explains, Gramsci's organic intellectual works "on two fronts at the same time"—the intellectual front and the more broadly civic one ("Cultural Studies and Its Theoretical" 268). We take this to indicate the importance of moving beyond academic critique to more broadly accessible arguments and political action.

The Birmingham school and some of its theoretical patrons (e.g., Foucault) were also influenced by structuralist strands of cultural studies, which Hall contrasts to the culturalist strand. Developed, in part, out of Althusser's neo-Marxist notion of ideology and Levi-Strauss's notion of culture as the "categories and frameworks in thought and language through which different societies classified out their existence" (Hall, "Cultural Studies: Two" 41), structuralism helped cultural studies better recognize "the intertextuality of texts in their institutional positions,... texts as sources of power,... textuality as a site of representation and resistance" (Hall, "Cultural Studies and Its Theoretical" 271). Structuralism's linguistic turn also focused cultural critique on the productive power of and struggles over ideology in texts and discursive formations.

More recent versions of structuralism (e.g., semiotics and deconstruction) have dominated U.S. cultural studies, especially in English departments. As Hall and several of this collection's contributors point out, these versions, partly through their institutionalization, tend to privilege academic critique over ethical action and to overlook the concrete, material exercises and connections of power (Hall, "Cultural Studies and Its Theoretical" 274).

Hall and other Birmingham school theorists have called for versions of cultural studies that draw on the best of both culturalist and structuralist insights and that recognize power as both discursive and material. Despite the continued dominance of semiotics in English studies, some U.S. critics have heeded this call. In their introduction to the collection *Cultural Studies*, Cary Nelson, Paula Treichler, and Grossberg offer a definition of culture as "the actual, grounded terrain of practices, representations, languages and customs of any specific historical society" as well as "the contradictory forms of 'common sense' which have taken root and helped to shape popular life" (5). In addition to referring to both material and discursive practices, here Nelson et al. refer to the ideological dimension of culture that is grounded in and helps to shape these practices. The essays in this collection work within this framing of culture, recognizing technical texts as connected to broader cultural practices, as always-already ideological, and as enmeshed in forms of power.

This incomplete overview of cultural studies has been pointing to our working definition, one that is flexible but not amorphous. Nelson, Treichler, and Grossberg describe cultural studies as a "bricolage," with its "choice of research practices depend[ing] upon the questions that are asked, and the questions depend[ing] on their context" (2). They also point out that cultural studies can't take just any form, however. As we're defining it, cultural studies involves critiquing and intervening in the conditions, circulation, and effects of discursive-material practices that are situated in concrete but dynamic sociohistorical formations, that participate in ideological struggles over knowledge legitimation, and that help shape identities. This definition emphasizes, then, the predispositions to account for technical communication's broader cultural conditions and power dynamics, to ethically critique its shifting functions and effects (especially subjective ones), and to intervene in hegemonic forms of power. Each contributor offers a different spin on this definition, of course, depending partly on the cultural studies tradition(s) on which he or she draws, but all contributors demonstrate cultural studies' transformative potential.

Technical Communication Scholarship at the Crossroads of Cultural Studies

In contrast to the relative abundance of work on cultural studies approaches to composition (for examples, see Berlin; Fitts and France), only a handful of technical communication scholars have tapped into cultural studies. That is why this collection is the first of its kind. Technical and professional communication scholars have been laying the groundwork for our collection for some time, however. Several have turned to classical rhetoric, radical pedagogy, feminist theory, and other critical traditions to emphasize the ethical, ideological, and political dimensions of technical communication. In two related articles, Steven B. Katz uses classical rhetoric to critique what he calls the "ethic of expediency" driving technical communication. Like Katz, Dale Sullivan argues that we must approach technical communication as phronesis (practical wisdom) as well as praxis (social action). As Sullivan points out, phronesis involves ethical deliberation about technical communication's effects, a concern shared by cultural studies (378). Indebted to radical educational theory, Herndl calls for research approaches and pedagogies that critique the broader relations of power inherent in professional communication and that enable ethical, public action based on this critique. Nancy Blyer takes up Herndl's invitation, developing a critical research perspective that borrows from feminist, radical educational, and participatory action methods to reshape the researcher-participant relationship in more egalitarian ways (33).

A few scholars have more directly drawn on critical cultural theory to account for technical communication's broader cultural conditions, power relations, and circulation. James E. Porter, Patricia Sullivan, Stuart Blythe, Jeffrey T. Grabill, and Libby Miles draw on cultural geography to develop a methodology for critiquing "institutions as rhetorical systems of decision making that exercise power through the design of space (both material and discursive)" (621). Here they expand and reconfigure our field's typical focus on studies of workplace culture within discrete organizations. In her seminal essay that begins section two of the collection, Bernadette Longo draws on Foucault to similarly move beyond uncritical, narrowly framed social constructionist research and advocate a cultural studies approach that critiques "how struggles for knowledge legitimation taking place within technical writing practices are influenced by institutional, political, economic, and/or social relationships, pressures, and tensions within cultural contexts that transcend any one affiliated group" (61–62). In *Spurious Coin*, Longo's cultural history of technical writing in twentieth-century U.S. institutions, she employs a primarily Foucaultian cultural studies frame to examine how technical writing participated in economies of scientific knowledge legitimation and in management control systems. Through this participation, Longo explains, technical writing served as a hegemonic tool for maintaining cultural and material capital and for stabilizing the "social distinction between people who have technical knowledge and those who do not" (3).

In the pedagogical arena, J. Blake Scott adapts Richard Johnson's notion of the cultural circuit to develop a heuristic that helps students critique and respond to the broader effects of their work as it circulates and is transformed ("Tracking"). Others have mobilized cultural studies articulation theory to review technical communication as an ongoing cultural process of creating meaning through linkages of power. In their often-cited article reprinted in this volume's first section, Jennifer Daryl Slack, David James Miller, and Jeffrey Doak argue that we should teach our students that they are more than transmitters or translators, but active contributors to an ongoing process of articulation. Henry similarly prompts his students to examine constructs of technical authorship and to wield this analysis to reposition themselves in more empowering workplace roles ("Teaching").

The recent issue of the *Journal of Business and Technical Communication* guest edited by Herndl and dedicated to "critical practice" contains several articles that "understand technical and professional communication as a cultural activity and as a medium for producing knowledge that is always politically interested" and that "generate really useful knowledge that opens up possibilities for action, however circumscribed or local" (Herndl, "Introduction" 3). This special issue powerfully demonstrates the promise of critical cultural theory for interventive political action as well as critique.

Imaginative and inspiring, the work of these scholars and other precursors to this collection's contributors has set the stage for more specific applications of cultural studies, developed here. This collection is the first of its kind, the first book project to collect work on the cultural studies–technical communication intersection, previously scattered across various forums. We hope it will further advance a shift in our field from the social turn to the cultural one.

Technical Communication's Hyperpragmatist Legacy

The main exigency for this collection is our alarm about the hyperpragmatist trajectory of our field. Despite attempts to challenge it, hyperpragmatism continues to dominate technical communication research and teaching, even coopting those practices that could be transformative (see Scott's chapter on service-learning). Before we elaborate on the limitations of hyperpragmatism, we want to trace the former's legacy in technical communication.

The earliest technical communication courses, essentially English for engineering courses, were mechanical but not particularly pragmatic. As Katherine Adams points out in her history of professional writing instruction, they were heavily influenced by the current traditional paradigm of first-year composition (137). This meant that they focused on mechanical correctness, clarity, and the modes of exposition and description. Carolyn Miller and Longo have traced this focus to a "pervasive positivist view of science" (Miller 610). This pedagogy was mostly antirhetorical. Gradually, technical writing courses began combining current traditional rhetoric with more vocational, practical concerns, adding topics and forms (e.g., report forms) used in science, engineering, and management.

Unlike first-year composition, which merged current traditional rhetoric with humanist literary study, early technical writing pedagogy was marked by a tension between these two approaches. Teresa Kynell describes this as a tension between utilitarianism and humanism. Kynell goes on to explain that advocates of the latter sometimes referred to the study of literature, history, and other humanist subjects as "culture studies" (10). This term referred to something quite different from most forms of cultural studies today, of course. Instead of cultural critique, this type was concerned with the refinement of taste, a marker of the bourgeois class. Its advocates saw such refinement as a way to elevate the social and professional status of engineers (e.g., into management positions) and the field of engineering (see Longo, *Spurious* 140). Even advocates of more utilitarian approaches saw technical writing as a way to maintain and extend class privilege.

For a time, utilitarian approaches were not just in tension with humanist ones but in tension with scientific ones as well. Influential textbook author T. A. Rickard, for example, argued that technical writing must be kept "pure" from the practical demands of engineering in order to safeguard the privileged status of scientific knowledge (Longo, *Spurious* 66). Rickard even argued that technical writing's contribution to the "general fund" of scientific knowledge was crucial to the betterment of humankind (Longo 63).

Technical writing became less the province of science, however, as it developed alongside engineering management systems. Longo explains: "As engineers designed management systems to make workers as efficient as the machines with which they worked, they also designed intricate technical communication systems as the mechanism for effecting operations control for maximum efficiency" (79).

As Kynell's title of *Writing in a Milieu of Utility* suggests, utilitarian, vocational concerns overrode humanist ones (as well as scientific ones) in most technical writing curricula, partly because of the persistent assessment of engineering graduates as poor writers. By the 1920s and 1930s, technical writing (i.e., English for engineering) courses were gradually becoming better suited to vocational needs, driven by the "engineer's need for . . . practical, real-world writing" (Kynell 68). According to other historians of technical writing, namely Robert Connors and Longo, most curricula were still largely formalist, largely revolving around the modes, clarity and the "plain style," and forms (see Longo, *Spurious* 146–47). At the same time, technical writing pedagogy included more report and letter forms along with basic attention to engineering audiences.

The 1940s and 1950s saw the establishment of the modern technical communication course, the move from engineering colleges to English departments, and the rise of technical communication as a field involved in research (we should note, too, that the discipline of composition began to emerge out of the communication skills movement during this same time period). World War II was the biggest influence on these developments, of course, as the technology boom during and after the war created a need for technical communication specialists. The new version of technical communication was more rhetorical but also more utilitarian than its predecessors. Audience and readability were given more presence, but typically in a fairly instrumental way; that is, more complex issues like ethics and the values and contexts of audiences were not addressed. Students were taught even more practical techniques and technical forms, such as the technical article, manual, and later, the proposal.

Longo notes that textbooks published just after World War II departed from their predecessors by straightforwardly interpellating technical writers as "individuals concerned with their own personal gain," a move that technical writing textbooks have continued to make ever since (75). Despite this stronger concern for personal economic success, the newly established field of technical writing saw its practitioners largely relegated to "lower-paid help to relieve higher-paid engineers and scientists" (Longo 144).

It was out of these historical developments that a more fully developed hyperpragmatism emerged. As we're defining it, hyperpragmatism is a hegemonic ideology and set of practices that privileges utilitarian efficiency and effectiveness, including rhetorical effectiveness, at the expense of sustained reflection, critique, or ethical action. In its more extreme forms, hyperpragmatism can be driven by an ethic of expediency, to use Katz's term. Katz argues that in our capitalist, highly technological society, expediency can become "both a means and an end," "a virtue itself" that subsumes all other ethical considerations ("Ethic" 270). The main goal of hyperpragmatist pedagogy is to ensure the technical writer's (and technical writing student's) professional assimilation and success as measured by vocational rather than more broadly civic terms.

In the last few decades of the past century, hyperpragmatism took more rhetorical and social forms as new theories were incorporated and technical writing was further institutionalized as a discipline. Early writing process theories were in many ways quite compatible with current traditional and other formalist ones, as Sharon Crowley explains, and therefore didn't transform composition pedagogy as much as our histories suggest they did (although they did help create the semester-long project course in technical communication) (211). Social process and social constructionist theory, influenced by pragmatism, did more to help the field move beyond purely utilitarian concerns with techne to include considerations of social praxis. These theories and pedagogies, which now seem commonplace, helped us view genres as social action, account for the conventions and values of discourse communities, and emphasize students' enculturation into these discourse communities. More teachers began using cases and real-world assignments that put students in workplace roles and presented them with concrete audiences and contexts. Numerous scholars began researching the rhetorical practices of various disciplinary and workplace discourse communities in order to better understand their dynamics and thereby help students more successfully conform to them.

Such developments are certainly improvements on more purely utilitarian or current traditional approaches, but they often stop short of enabling cultural critique and ethical intervention. Like their more utilitarian precursors, their rhetorical dimension can be narrowly vocational, and their main goals are to help students better understand, conform to, and succeed in their disciplinary and workplace discourse communities.

Although pragmatism and social constructionism mark an epistemological break from current traditional rhetoric, they don't mark a political break from it. These theories pretend to be apolitical but, like current traditional rhetoric, are largely conservative in that they value accommodation and conformity (to conventions, practices, and values) ahead of critique. Even work that touts innovation along with accommodation often narrowly defines this innovation in terms of rhetorical and organizational effectiveness and productivity (see, for example, Spilka 209). Newer forms of hyperpragmatism are not only conservative but also liberal in their emphasis on the student's professional formation and goals. As Crowley notes, liberal approaches see their primary aim as helping individuals get better at whatever they want to do (219). Whatever form it takes, hyperpragmatism is decidedly not radical.

The transformative potential of more robust rhetorical/social approaches can be squelched all too easily by hyperpragmatism. The corporatization of the university—including the move toward more (economically) efficient pedagogical models and the growing funding and curricular ties of technical communication programs to industry—can work to squelch critique on the institutional level. As Jack Bushnell points out, we have become "training departments for corporate 'clients' who provide us with internships and fellowships . . . and ever increasing numbers of good-paying jobs" for our students (175–76). This corporatization has, not surprisingly, shaped our students' attitudes about and expectations of their education, which can themselves be powerful forces for hyperpragmatism. Our own professional advancement is in many cases tied to how well we meet hyperpragmatist expectations of our students, administrators, corporate partners, and the larger public.

Limitations of Hyperpragmatism, Contributions of Cultural Studies

It is tempting, we admit, to succumb to the cultural pressures that keep hyperpragmatism in place, and we hold some pragmatist values ourselves (e.g., we want our students to get good jobs, etc.). But we believe hyperpragmatism—even in its more robustly rhetorical, social forms—can be limiting in several key ways. We are not the first to point these out, but we think that collectively they point to the need for cultural studies. As the chapters in this collection demonstrate, cultural studies holds much promise for correcting the following limitations of hyperpragmatism.

First, hyperpragmatism overlooks the broader web of conditions, relations, and power dynamics of which technical communication is part. Longo and Herndl critique dominant technical and professional communication research for its narrow focus on the production processes of discrete discourse communities. The limitation of this research, in the words of Herndl, is that it "describes the production of meaning but not the social, political, and economic sources of power which authorize this production or the cultural work such discourse performs" (351). Hyperpragmatist pedagogy follows the lead of this research, typically focusing students' invention on practical elements of textual production. Common invention heuristics such as the rhetorical triangle or forum analysis often encourage students to examine their immediate rhetorical situations and discourse communities but not look further into the cross-cultural and postproduction trajectories of their work.

Reviewing technical communication as more broadly cultural first entails accounting for its broader, shifting conditions of possibility. The cultural studies notion of articulation can be useful here, as Slack et al. demonstrate. Citing Hall, Slack et al. explain articulation as the ongoing process by which coherent cultural forms (e.g., technical texts) are produced out of nonnecessary linkages of various cultural elements, including ideologies and material forms (25, 28). Another concept that can extend our thinking about technical communication is Richard Johnson's notion of the cultural circuit, which emphasizes the broader circulation and transformations of technical communication and its effects as it is produced, textually embodied, distributed, regulated, consumed, and integrated into lived experience.

Hyperpragmatism can also look past the regulatory power that conditions (and is reinforced by) technical communication, partly by pretending to be apolitical. Guided by a still lingering positivism, some of our field's research and pedagogy continues to treat technical communication as somehow separate from the political and ideological dimensions of culture, a demarcation Jameson warns against:

> the convenient working distinction between cultural texts that are . . . political and those that are not becomes something worse than an error: namely, a symptom and a reinforcement of the reification and privatization of contemporary life. Such a distinction reconfirms that structural, experiential, and conceptual gap between . . . the political and the poetic [or technical] . . . which . . . alienates us from our speech itself. (20)

When we view technical texts as apolitical and nonideological, Jameson suggests, we misunderstand them and limit our ability to transform them for civic good.

Even some social constructionist research, despite moving beyond positivism, avoids questions about the politics of knowledge legitimation and the exclusionary effects of power (Longo, "Approach" 54). Pedagogy based on such research, adds Herndl, "will produce students . . . who cannot perceive the cultural consequences of a dominant discourse or the alternate understandings it excludes" (350). Recognizing the social dimension of technical communication is a starting point but can stop short of understanding technical communication as part of power/knowledge formations that include ideologies, institutional constraints, economic pressures, and other cultural forces.

The ideologies that regulate technical communication can be difficult to recognize when these networks of interpretations function as defaults, as what Vincent Leitch calls "regimes of reason." Leitch explains that "the conscious and unconscious oppositional and ruling values within social formations, however contradictorily compacted, comprise 'regimes of reason' or of 'unreason,' as the case may be" (1). Because regimes of reason seem commonsensical, they often go unnoticed. An important goal of cultural studies, and an important goal of the essays in this volume, is to make common sense about technical communication uncommon again in order to interrogate its values, functions, and effects.

Cultural studies approaches can help us and our students review technical communication as regulated by and enacted as power. As Grossberg explains, "Cultural studies is always interested in how power infiltrates, contaminates, limits, and empowers the possibilities that people have to live their lives in dignified and secure ways" (257). Although the approaches we advocate follow Foucault in resisting totalizing and repressive notions of power, instead viewing it as disbursed and productive, they do recognize that power can disable as well as enable, exclude as well as include, delegitimate knowledge as well as legitimate it. Cultural studies can help us as researchers and teachers reframe pragmatic questions about how to reproduce knowledge with more Foucaultian questions about why only certain knowledges are legitimated and to whose benefit. Longo recommends starting with Foucault's basic archaeological question: "How is it that one particular statement appeared rather than another?" (27; cited in Longo 120; 62 in original).

In addition to the power dynamics of the technical communication that we and our students produce, cultural studies can help us be more aware of and responsive to the power dynamics of our research and teaching practices. Cultural studies is nothing if not self-reflexive about its goals and methods. In *Opening Spaces*, Sullivan and Porter develop several ways of operationalizing this reflexivity. They suggest that technical and professional communication researchers map their positionality in relation to others in research scene maps, for example, and make their agendas explicit in advocacy charts. Some of these same strategies could be undertaken by teachers, students, and practitioners as well.

Yet another related limitation of hyperpragmatism is its explanatory rather than critical stance, a stance driven by the goal of accommodation rather than transformation. The goals of hyperpragmatism are conformity, expediency, and success, narrowly defined. These goals can co-opt as well as foreclose critique, as Henry and Katz point out, replacing ethical standards with those of economic expediency. Henry laments that "quality writing" is too often defined as "writing that maximizes investors' returns on investments, and one can imagine writerly sensibility being shaped to this end, if only to maintain one's current job as long as possible." "Eliminated from the equation," he adds, "are issues of ethics, of workers' interrelationships with colleagues, of the quality of life in the local work culture, and of the ultimate effects on other populations of the writing in which one is engaged" (8). Even when ethics are involved, explains Katz, they "can be based as much on realpolitick and power as they are ideals of participation and pluralism" ("Aristotle's" 50). Katz challenges us to make sure that the phronesis or ethical judgment that we teach students is based on more than narrow corporate values.

Cultural studies can push our research, pedagogy, and practice to critically assess and problematize the hegemonic values and functions of technical communication. Instead of only seeking to explain technical communication, we should evaluate the ethics of its functions and effects, asking such questions as "Whose values does technical communication privilege?" "Who is included and who is excluded by these practices and how?" "Who benefits and who loses?" and "How are these practices beneficial and/or harmful?"

Johnson, Grossberg, and other critics assert that cultural studies is particularly concerned with assessing subject-related effects of power. In this respect we might critique the ways technical communication positions and interpellates users and the ways it helps shape their lived experience. We might also critique, as Henry has his students do, how the identities and work lives of technical communicators are institutionally and socially constructed. Greg Wilson similarly exhorts us to help students develop more expansive views of their agency, to "think differently about the relationships between technical concepts and to critique their relationships as communicators and social actors to technology and authority" (74).

Although researchers and teachers may not always agree on the ethical principles guiding their critiques, we echo Sullivan and Porter's belief that these goals should begin with respecting difference, being attentive and responsive to others, creating more egalitarian and just practices, and empowering those affected by our work (110). Others, too, namely Blyler, Robert Johnson, and Michael J. Salvo, have similarly argued for technical communication practices that are more inclusive of and empowering to users and others affected by them.

Along with devaluing critique, hyperpragmatism devalues ethical intervention. Many technical communication teacher-scholars, practitioners, and students are content with reproducing the status quo rather than revising it. Henry Giroux cynically argues that corporate culture portrays citizenship "as an utterly privatized affair whose aim is to produce competitive self-interested individuals vying for their own material and ideological gain" (30). We too rarely take on the role of citizen-advocate in our own research and writing, and we rarely ask students to take on such a role. Even when we push students toward civic critique, we often don't encourage them to develop plans for enacting this critique through policy changes and other initiatives. As Herndl points out, most technical communication courses spend "relatively little energy analyzing the modes and possibilities for dissent, resistance, and revision" (349). Instead, students spend their time accommodating "consumers" or "users" of specific products or services.

In reproducing hegemonic power structures, from the regulatory mechanisms of scientific institutions to the management practices of transnational corporations, hyperpragmatic technical communication runs the risk of disempowering or even harming many, including technical communicators themselves. Herndl and Dale Sullivan have critiqued technical communication research and pedagogy, respectively, on this count, Sullivan asking "if we enculturate students in the technical writing classroom, at least in part by teaching technical genres that reinforce the dominance of the technological system, how can we then call them to responsible social action?" (377). Many of our students may go on to create discourse for industries (e.g., financial consulting, insurance, pharmaceutical, energy, and high-tech weapons industries) that in some cases exploit, dismiss the needs of, or threaten to directly harm groups of people. If trained only in hyperpragmatist approaches, however, these students might find it difficult to ethically respond to problematic functions and effects of the discourses in which they're implicated, especially given that current corporate culture often positions them and their expertise as "marginal in the organization's life" (Henry 9).

Once again, cultural studies offers a possible correction. Perhaps the most important function of cultural studies is to translate critique into ethical civic action. As Foucault illustrated in his life and built into his notion of genealogy, critical intellectual work can inform tactical interventions in ongoing symbolic and material struggles. De Certeau also advocated the subversion of dominant knowledge systems by engaging in unsanctioned, unofficial, tactical activities. Like Gramsci and Foucault, de Certeau held the hope that even subjugated groups of people help transform culture over time through their actions:

> The purpose of this [unsanctioned] work is to make explicit the systems of operational combination . . . which also compose a "culture," and to bring to light the models of action characteristic of users whose status as the dominated element in a society (a status that does not mean that they are either passive or docile) is concealed by the euphemistic term "consumers." Everyday life invents itself by poaching in countless ways on the property of others. (xi-xii)

Part of de Certeau's message here is that culture does not simply render us docile, but we respond to and transform culture. Sometimes cultural transformations can be abrupt and large-scale, and such events provide rich sites for study. Yet cultural studies can also focus on ongoing, local struggles over knowledge-power structures in and across institutions, organizations, and daily life. The studies in this collection take up such struggles at a variety of sites, including government agencies, corporations, community initiatives, academic programs, and individual classrooms. They show that far from being a negative or simply deconstructive enterprise, critical cultural study can work to restore the ethical promise of technical communication, too often squelched by economically and technologically expedient elements of our culture.

The chapters in this collection point to a more visible cultural turn in our field's trajectory. Although they do not all draw from the same theories or traditions of cultural studies or apply them in the same ways, their divergences attest to the richness of cultural studies and the expansive potential for its cross-fertilization with technical communication. Taken together, the chapters offer a broad but certainly not complete repertoire of approaches. Despite their differences, this collection's chapters all demonstrate the basic cultural studies moves that we outlined earlier: situating technical communication in concrete but dynamic sociohistorical formations; accounting for technical communication's broader cultural conditions, relations, and effects; self-reflexively critiquing technical communication's functions in ideological, power-laden struggles over knowledge; laying the groundwork for ethically intervening in disempowering or otherwise harmful practices.

Demonstrating the cross-fertilization of cultural studies and technical communication is quite different than offering the former as a panacea for the latter. The collection's chapters are careful to explain the potential pitfalls of cultural studies applications, to adapt cultural studies approaches to technical communication concerns, and to show how technical communication research can inform cultural studies. Beverly Sauer, for example, explains how the cultural studies impulse to radically contextualize may not always work in the interest of workers engaged in risk communication. Just as compositionists' applications of liberatory pedagogy have fed back into and informed this movement, we hope that the applications offered here can inform the larger enterprise of cultural studies.

We have divided the collection into three (overlapping) sections—theory, research, and pedagogy—based on the primary emphasis of the chapters that fall under them. The chapters in the theory section complicate and reconceptualize common technical communication concepts and practices—such as usability and email—through a cultural studies lens. The chapters in the research section critique, refocus, and extend methods for studying various types of technical communication. The chapters in the pedagogy section explore ways that cultural studies principles can enhance and redirect curricula and pedagogical approaches. These are somewhat arbitrary distinctions, we admit, as the three categories and their chapters necessarily overlap. The theory chapters have clear implications for technical communication practice, research methods, and pedagogy, for example, and the research methods and pedagogical approaches certainly inform each other.

In addition to the original essays, we have begun each section with an already-published essay. These reprinted chapters are among the field's first and most important explorations into the cultural studies-technical communication intersection. As such, they have helped to initiate the conversation that the other chapters join. We offer this collection of chapters not as a set of models but as a tribute to the earlier critical work by Herndl, Katz, Dale Sullivan, and others, and as a dialogue that practitioners, scholars, teachers, and students alike can build on, learn from, and even challenge. As Nelson, Treichler, and Grossberg point out, cultural studies is not and should never be a static, homogeneous enterprise. We therefore hope that this collection will spark a dynamic conversation that explores still more pathways to and from the cultural studies-technical communication intersection.

NOTE

1. In this collection, both Britt and Grabill ground their arguments in theories from critical anthropology and sociology. In his groundbreaking essay critiquing professional writing research and pedagogy, Herndl draws largely on the tradition of critical or liberatory pedagogy. Demonstrating yet another cultural studies approach, Longaker uses neo-Marxist, macroeconomic theory to call our attention back to the broader capitalist contexts of technical communication practices.

WORKS CITED

Adams, Katherine. *A History of Professional Writing Instruction in American Colleges: Years of Acceptance, Growth, and Doubt.* Dallas: Southern Methodist UP, 1993.

Berlin, James A. *Rhetorics, Poetics, and Cultures: Refiguring College English Studies.* 2nd ed. West Lafayette, IN: Parlor P, 2003.

Blyler, Nancy. "Taking a Political Turn: The Critical Perspective and Research in Professional Communication." *Technical Communication Quarterly* 7.1 (1998): 33–52.

Crowley, Sharon. *Composition in the University: Historical and Polemical Essays.* Pittsburgh: U of Pittsburgh P, 1998.

de Certeau, Michel. *The Practice of Everyday Life.* Trans. Steven Rendall. 1984. Berkeley: U of California P, 1988.

During, Simon. "Introduction." *The Cultural Studies Reader.* Ed. Simon During. London: Routledge, 1993. 1–28.

Fitts, Karen, and Alan W. France, eds. *Left Margins: Cultural Studies and Composition Pedagogy.* Albany, NY: SUNY P, 1995.

Foucault, Michel. *Archeology of Knowledge and the Discourse on Language.* Trans. A. M. Sheridan Smith. 1969. New York: Barnes and Noble, 1972.

Gramsci, Antonio. *Selections from the Prison Notebooks.* Ed. and trans. Quintin Hoare and Geoffrey Nowell Smith. New York: International Publishers, 1971.

Grossberg, Lawrence. *Bringing It All Back Home: Essay on Cultural Studies*. Durham, NC: Duke UP, 1997.

Hall, Stuart. "Cultural Studies and Its Theoretical Legacies." *Stuart Hall: Critical Dialogues in Cultural Studies*. Ed. David Morley and Kuan-Hsing Chen. New York: Routledge, 1996. 263–275.

———. "Cultural Studies: Two Paradigms." *Media, Culture and Society* 2.1 (1980): 57-72.

Henry, Jim. "Writing Workplace Cultures." *College Composition and Communication Online* 53.2 (2001): www.ncte.org/ccc/2/53.2/henry/article.html.

Herndl, Carl G. "Introduction: The Legacy of Critique and the Promise of Practice." *Journal of Business and Technical Communication* 18.1 (2004): 3–8.

———. "Teaching Discourse and Reproducing Culture: A Critique of Research and Pedagogy in Professional and Non-Academic Writing." *College Composition and Communication* 44 (1993): 349–363.

Herndl, Carl G., and Cynthia A. Nahrwold. "Research as Social Practice." *Written Communication* 17.2 (April 2000): 258–296.

Jameson, Fredric. *The Political Unconscious: Narrative as a Socially Symbolic Act*. Ithaca: Cornell UP, 1981.

Johnson, Richard. "What Is Cultural Studies Anyway?" *Social Text* 6 (1987): 38–80.

Johnson, Robert R. *User-Centered Technology: A Rhetorical Theory for Computers and Other Mundane Artifacts*. Albany: SUNY Press, 1998.

Katz, Steven B. "Aristotle's Rhetoric, Hitler's Program, and the Ideological Problem of Praxis, Power, and Professional Discourse." *Journal of Business and Technical Communication* 7.1 (1993): 37–62.

———. "The Ethic of Expediency: Classical Rhetoric, Technology, and the Holocaust." *College English* 54.3 (March 1992): 255–275.

Kynell, Teresa C. *Writing in a Milieu of Utility: The Move to Technical Communication in American Engineering Programs, 1850–1950*. Norwood, NJ: Ablex, 1996.

Leitch, Vincent B. *Cultural Criticism, Literary Theory, Poststructuralism*. New York: Columbia UP, 1992.

Longaker, Mark Garrett. "Back to Basics: An Apology for Economism in Technical Writing Scholarship." *Technical Commication Quarterly* 15.1 (2006): 9–29.

Longo, Bernadette. "An Approach for Applying Cultural Study Theory to Technical Writing Research." *Technical Communication Quarterly* 7 (1998): 53–73.

———. *Spurious Coin: A History of Science, Management, and Technical Writing*. Albany: SUNY P, 2000.

Miller, Carolyn R. "A Humanistic Rationale for Technical Writing." *College English* 40 (1979): 610-617.

Nelson, Cary, Paula A. Treichler, and Lawrence Grossberg. "Cultural Studies: A User's Guide to This Book." *Cultural Studies.* Ed. Lawrence Grossberg, Cary Nelson, and Paula A. Treichler. New York: Routledge, 1992. 1–16.

Porter, James E., Patricia Sullivan, Stuart Blythe, Jeffrey T. Grabill, and Libby Miles. "Institutional Critique: A Rhetorical Methodology for Change." *College Composition and Communication* 51.4 (June 2000): 610–642.

Salvo, Michael J. "Ethics of Engagement: User-Centered Design and Rhetorical Methodology." *Technical Communication Quarterly* 10.3 (2001): 273–290.

Scott, J. Blake. *Risky Rhetoric: AIDS and the Cultural Practices of HIV Testing.* Carbondale, IL: Southern Illinois UP, 2003.

———. "Tracking Rapid HIV Testing Through the Cultural Circuit: Implications for Technical Communication." *Journal of Business and Technical Communication* 18.2 (2004): 198–219.

Slack, Jennifer Daryl, David James Miller, and Jeffrey Doak. "The Technical Communicator as Author: Meaning, Power, Authority." *Journal of Business and Technical Communication* 7.1 (1993): 12–36.

Spilka, Rachel. "Influencing Workplace Practice: A Challenge for Professional Writing Specialists in Academia." *Writing in the Workplace: New Research Perspectives.* Ed. Rachel Spilka. Carbondale, IL: Southern Illinois UP, 1993. 207–219.

Sullivan, Dale L. "Political-Ethical Implications of Defining Technical Communication as a Practice." *Journal of Advanced Composition* 10.2 (Fall 1990): 375–386.

Sullivan, Patricia, and James E. Porter. *Opening Spaces: Writing Technologies and Critical Research Practices.* Greenwich, CT: Ablex, 1997.

Williams, Raymond. *Culture and Society: 1780–1950.* 1958. New York: Columbia UP, 1983.

———. *Keywords: A Vocabulary of Culture and Society.* New York: Oxford UP, 1976.

Wilson, Greg. "Technical Communication and Late Capitalism: Considering a Postmodern Technical Communication Pedagogy." *Journal of Business and Technical Communication* 15.1 (2001): 72-99.

Theory

Since the nineteenth century, technical communication practices have most often been understood as functional and instrumental—means to ends. The means are clear communication; the ends are scientific facts. In this view of technical communication, theory is largely unarticulated, implicit, and unnecessary. Technique is sufficient.

In recent decades, however, a growing community of technical communication researchers and scholars has come to understand that practice is never theory-free and that being able to articulate a theoretical basis for practice is necessary for self-aware, professional development. In their 1993 article "The Technical Communicator as Author: Meaning Power, and Authority," Jennifer Daryl Slack, David James Miller, and Jeffrey Doak were on the forefront of cultural studies in technical communication, arguing that technical communicators are not merely neutral conduits for factual information, but are engaged in relations of power within cultural contexts. These authors first looked to Foucault's writing to question why some documents have authors, while others have writers, signatories, underwriters, or are left unattributed. Technical documents, naturally, do not have authors, or even writers. Technical writers are not granted authorship within our cultural contexts because the types of documents we write are not "authored" genres. Thus, the genre itself precludes authorship.

Given this historical situation, Slack et al. explain the development of technical communication practice and its cultural implications by reviewing three models (or views) of how meaning is made through technical communication:

- Transmission—meaning is conveyed from sender to receiver by a purveyor
- Translation—meaning is interpreted and reinterpreted by a mediator
- Articulation—meaning is articulated and rearticulated in cultural struggles by an author

After reviewing the positivist transmission model, the authors apply Stuart Hall's theory of encoding and decoding to describe the translation model in which messages participate in a "circuit of meaning production" (33; 177 in original). Finding that the communicator's role as mediator opens possibilities for authorial power, the authors argue that this role is "undertheorized" (37; 181 in original) unless the author has agency for language production within a cultural context. Their articulation model "asserts that any identity in the social formation must be understood as the nonnecessary connection between elements that constitute it" (37; 182 in original). And further, they argue that "identity is culturally... struggled over in ongoing processes of disarticulation and rearticulation" (39; 183 in original). They call on Lawrence Grossberg's theory of power as the agency to empower or disempower possibilities or meanings, concluding that once technical communicators realize that we have power to create some meanings and not to create other possible meanings, we also must be held responsible for our creations.

Slack, Miller, and Doak carefully trace the transformation of the technical communicator's role from neutral conduit to implicated translator to responsible author. In laying out this evolution, the authors explain the rationale for why cultural critique is important and relevant to research in technical communication: cultural critique lets us talk about our social responsibilities as communication professionals. We are still a young and somewhat undertheorized discipline. Yet our practices serve to (de)stabilize important rational and scientific knowledge/power structures in our culture. Only when technical communicators accept responsibility as authors within our cultural context can we begin to understand and control our practices and the technologies in which we are complicit.

In his article "Extreme Usability and Technical Communication," Bradley Dilger interrogates the technological ethic of "ease" and how this ethic plays out in current notions of extreme usability. Dilger peels the onion of usability, so to speak, examining each layer for the theoretical underpinnings that valorize simplicity, ease of use, pragmatism, and commercialism. He places these notions within a historical context of ergonomics and human factors engineering, "which emerged... immediately after World War II, in response to the need to make complex military equipment less difficult to operate" (50). This military exigency combined with postwar consumer culture and the concept of usable consumer products emerged as the forebear of its application to computer technologies. By placing usability practices within their historic and cultural contexts, Dilger argues that traditional usability seeks to educate naive user know-how (in de Certeau's terms) into the rationalized technical knowledge and practices that support the user's successful consumption of commercial products. Rather than being merely translators of technology or mediators between technologies and people, Dilger sees technical communicators as "authors" (in Slack et al.'s terms): proponents of commercial and industrial goals and the capitalist ideology that underpins them. In these terms, technical communicators need to understand that when we promote an ethic of "ease," we are sanctioning a machine ideology that Steven B. Katz described as an "ethic of expediency" in his earlier work "The Ethic of Expedience: Classical Rhetoric, Technology, and the Holocaust."

Dilger concludes that "fully developed models of usability" (65) are needed to more clearly understand the implications of valorizing "ease" and what happens to people when their practices and knowledge are shaped to this technological imperative. While he calls for more fully articulated and complex theories of product usability and relationships between people and products, Dilger also begins to articulate such a theory that places these objects of inquiry within their cultural and historic contexts.

Myra G. Moses and Steven B. Katz turn their critical gaze to email as another naturalized cultural phenomenon in "The Phantom Machine: The Invisible Ideology of Email." These authors argue that email is a commercial product that embodies the technological imperatives of productivity, efficiency, and speed. Like Dilger, they begin by examining some consequences of imposing this technological ethic on people in the public spaces of their lives: workplaces, commercial spaces, and so on. What Moses and Katz ultimately uncover, though, is email's power to blur public and private spaces. For example, business and personal messages flood email inboxes in ever increasing numbers. While corporate decision-makers may have frowned on personal telephone calls in earlier eras, today's managers tolerate personal email because the technology has led to high levels of increased worker productivity. Yet this "benefit" has troubling ramifications for people who now have unclear boundaries between work and personal lives, as well as work and personal messages.

Moses and Katz thoroughly trace how the rise of email has strengthened our willingness to "buy into" technological imperatives in our personal lives, leading us to expect a means-ends model of personal relationships that mirrors the productivity ethic-driving work relationships. Their interrogation of the ideological underpinnings and outcomes of adopting a machine-based, commercial ethic in our personal lives again places technical communicators at the center of a cultural power system based on rationalized, scientific knowledge. Following the lead of Slack, Miller, and Doak, Moses and Katz articulate how technical communication practices go beyond merely translating our behaviors. These practices encourage us to adapt our behaviors, our personal relationships and ultimately our worldviews to technological imperatives. Moses and Katz conclude,

"Ideologically, work and leisure have become virtually interchangeable" (75) and "[r]elations—both personal and business—become focused on the technical production of communication" (96).

All three of these studies foreground the theoretical underpinnings of their explorations, showing how cultural theories can illuminate relationships between people and technologies, between knowledge-making and cultural power, between technological development and social responsibility. Each of these studies explicitly discusses the theoretical foundation that allowed the authors to illustrate how technical communicators participate in cultural work through the choices we make, the language we privilege, and the masters we serve. They begin to construct theories of usability and technological accommodation that can be applied to other objects of inquiry within the technical communication field. These articles serve as strong examples of the power to affect cultural change that technical communicators already possess through our use and control of language. They also show us how cultural critique can help us claim this power and use it more intentionally through our research and professional practices.

Chapter One

The Technical Communicator as Author

Meaning, Power, Authority

Jennifer Daryl Slack, David James Miller, and Jeffrey Doak

In his essay, "What Is an Author?" Michel Foucault observes that

> in our culture, the name of an author is a variable that accompanies only certain texts to the exclusion of others: a private letter may have a signatory, but it does not have an author; a contract can have an underwriter, but it does not have an author; and, similarly, an anonymous poster attached to a wall may have a writer, but he cannot be an author (124).

From this, Foucault concludes that "the function of an author is to characterize the existence, circulation, and operation of certain discourses within a society" (124). At its most mundane, this is simply to note the fact that certain discourses are granted the privilege of authorship while others are denied this privilege. It is more remarkable to notice, with Foucault, that this very fact suggests an inversion of the way in which we typically understand the relation between an author and a discourse: Rather than authors producing certain discourses, certain discourses are understood to produce authors. To grant authorship to a discourse is to grant that discourse a certain authority. In a peculiar turn of events, this authority comes to reside in the author, the author produced by the discourse itself. Thus it becomes evident that authorship is a manner of valorizing certain discourses over against others. As such, authorship empowers certain individuals while at the same time renders transparent the contributions of others.

The discourses created by technical communicators have not been considered authored discourses; the technical communicator may be a transmitter of messages or a translator of meanings, but he or she is not—or at least

---

This chapter originally appeared in the *Journal of Business and Technical Communication* 7.1 (1993): 12–36. Reprinted with permission of SAGE Publications, Inc.

not until now—considered to be an author. We have come to see that technical communicators, as well as other professional communicators, are engaged in the process of what Marilyn Cooper has called *participatory communication.* In "Model(s) for Educating Professional Communicators," Cooper writes:

> I am defining communication as participatory communication and the role of...communicators as one of...working together to create common interests, to construct the ideals of our society, [and in light of these ideals] to examine the ends of [our] action. Professionals who communicate should be involved in this endeavor too....It is [at least part] of the function of professional communicators—whether they know it or not. (12)

THE RELEVANCE OF COMMUNICATION THEORY

There are striking parallels to be found by comparing descriptions of the technical communicator (descriptions and redescriptions of the role, task, and ethos of that communicator) with the progressive development of our theoretical understanding of the communication process itself. The most remarkable of these parallels may well lie in the emerging evidence of a symmetry between disparate images of the technical communicator and distinct—although ultimately interrelated—models of communication. What we propose is that, by comparing different images of the technical communicator with parallel developments in the study of communication, a new theoretical and practical image of that communicator—the technical communicator as author—can begin to be established. Reflecting on the historical development of communication theory over the course of the past ten years, scholars in communication have come to acknowledge that, at least with respect to the study of mass communication, two basic models of communication have gained ascendancy and, although this is less widely acknowledged, that a third is now gaining ground (see, for example, Fiske, *Introduction*; Carey). For our purposes, it is more useful to speak of these models not as models per se but as distinct views of communication. This is the case because, at bottom, each of these models seeks to express the morphology common to a collection of theories that otherwise appear more or less disparate. In this regard, the term "model" is misleading. It appears to set one theory of communication over against other such theories rather than gathering a number of specific theories together in a general conceptual classification. We have no interest in a general conceptual classification. We have no interest here in pitting one theory of communication against another. We are concerned with what these views, together, can teach us about the place of the technical communicator.

The first of these views—what we will refer to as the transmission view of communication—can be delimited in terms of a concern, for the most part, with the possibilities and problems involved in message transmission, that is, in conveying meaning from one point to another. The second—what we will call the translation view of communication—can be understood in terms of a primary concern with the constitution of meaning in the interpretation and reinterpretation of messages. The third—what we will call the articulation view of communication—can be grasped as a concern principally with the ongoing struggle to articulate and rearticulate meaning. With respect to each of these views of communication, the place of the technical communicator is located differently. In the first, the transmission view of communication, the technical communicator is a purveyor of meanings; in the second, the translation view of communication, the technical communicator is a mediator of meanings; in the third, the articulation view of communication, the technical communicator is an author who among others participates in articulating and rearticulating meanings.

Corresponding to variations in the place of the technical communicator as purveyor, mediator, or articulator of meanings, the place of the technical communicator—and of technical discourse itself—shifts in different relations of power. In the transmission view, the technical communicator remains the neutral vehicle facilitating the exercise of power. In the translation view, the technical communicator works to create symmetry within the negotiation of differential relations of power between sender and receiver. In the articulation view, the technical communicator is complicit in an ongoing articulation and rearticulation of relations of power. Ultimately, looking through the lens of articulation—as we do in this article—the different locations of the technical communicator implicate one another. That is, the technical communicator and technical discourse purvey, mediate, and articulate meaning. Likewise, the technical communicator and technical discourse facilitate, sustain, generate, and disrupt relations of power. But only by looking through the lens of articulation can we rearticulate the technical communicator and technical discourse as participating fully in the articulation of meaning and thereby fully empower the discourse as authorial.

Changing Conceptions of Meaning and Power: Transmission

Of the three views of communication, the transmission view has been the most clearly delineated. It has been extensively critiqued and often maligned such that it is nearly requisite to begin any introductory text on communication theory with an explanation and rejection of it. For the most part, contemporary communication theories are proposed in contradistinction to it. There are, consequently, many different versions of the position and ongoing disagreements about its precise historical and theoretical contours (see, e.g., "Ferment in the Field"). In general, however, the transmission view combines three defining characteristics:

1. the conception of communication as the transportation of messages;
2. the conception of the message—the meaning encoded by a sender and decoded by a receiver—as a measurable entity transmitted from one point to another by means of a clearly delineated channel; and
3. the conception of power as the power of the sender to effect, by means of this message, a desired mental and/or behavioral change in the receiver. This power is the power of the sender over the receiver.

The term "communication" has its origins in the concept of transportation (Williams; *Oxford English Dictionary* [*OED*]). Communications were the paths of transportation by means of which people at the centers of power could exercise control over those in the peripheries. The ability to move messages in a timely fashion across space by means of such communications was a necessary condition for political, economic, and religious domination. The emphasis in the historical development of new technologies of communication (from walkers, runners, horses, smoke signals, semaphore, print, telegraph, telephone, television, satellites, computers, fax machines, etc.) has been the transmission of knowledge and information in such a way as to *exercise control over space and people faster and farther*.

The implications for how meaning has been understood in communication theory are made clear by examining how communication as transportation gets tied to a theory of transmission. The work of Shannon and Weaver can be credited as a principal determinant in the shaping of such a view. Largely mathematical in character, Shannon and Weaver's conception of communication is as an explicitly linear form: The sender wishes to transmit meaning, but to do so it must be encoded in the form of a message (Shannon and Weaver called this *information*). The message is sent over a channel to a receiver who then decodes it to get out of it the meaning that was encoded. The process, when perfectly executed, results in the receiver's decoding exactly the same message that the sender intended to encode.

This basic model has been amended and elaborated on extensively (see, for example, Fiske, *Introduction*; McQuail and Windahl), but its orientation to meaning remains essentially the same. Meaning is something that is "packaged up" by the sender, shipped out, and "unwrapped" by the receiver, who can then act or think accordingly. Of course, there are numerous points in the process where difficulties can render the transmission less than perfect. The sender may encode the message poorly such that the message fails to contain the intended meaning. The decoder may decode poorly, not reading the intended meaning properly. There may also be "noise" in the channel that distorts the message so that, consequently, the meaning it contains is not received in the form in which it was sent. (Noise may take many forms, from static on the telephone line to the wandering mind of the listener during the transmission.)

In the transmission view of communication, meaning is a fixed entity; it moves in space "whole cloth" from origin to destination. Communication is successful when the meaning intended by the sender is received accurately, where accuracy is measured by comparing the desired response to the message with the actual response. Communication fails when these responses diverge. In the case of failure, the communicator must locate and correct the source of the failure in the process of encoding or in the noise of the transmission. Power is simply that which is exercised when the communication is successful. The sender has power when the receiver behaves in the intended manner. Power, like meaning, is something that can be possessed and measured; its measure is to be found in the response of the receiver.

Such a view of communication appears to dominate the early stages of the theory and practice of technical communication as it emerged within the college curricula of engineering schools. Based on research done by Robert J. Connors, we would characterize this phase in technical communication as dominating the field from the late 1800s until the 1950s but persisting into the present. In this phase, technical writing and engineering writing are treated as synonymous, and the task of the technical writing course is to teach engineers or their surrogates to encode the engineers' ideas (meanings) accurately and to provide a clear channel for transmission.

Technical writing courses developed this way in response to a series of changes in the practice of engineering and the development of the engineering curriculum. As engineering and its curriculum became more technically specialized and less humanistic during the period of rapid industrialization that followed the Civil War, complaints about the unbalanced education of students in technical schools mounted. Among other deficiencies, engineers, it was claimed, "couldn't write." To correct this imbalance, courses in engineering English (later technical writing; later still, technical communication) were developed. As Connors points out, by this time the notion of the "two cultures" split was so firmly in place that, as we would put it, the kind of meanings that required encoding were sufficiently different to warrant a completely different kind of English course (331). Engineering English courses were designed, among other things, to teach students to encode the special meanings of engineers.

Education in this phase has two components: the education of engineers and the education of surrogate engineers. Both are firmly anchored in a transmission view. The earliest, but again still persistent, effort to inculcate the skill of technical writing is to teach the engineer—as sender—to be a better encoder through the use of proper language, grammar, and style. Through such training, the intent is that engineers will learn to encode messages such that they will match their intentions. Further, in teaching engineers to transmit those properly coded messages using the proper forms, the intent is to ensure that the proper channels are chosen and that the transmission is sent with minimal noise. In James Souther's review of the evolution of technical writing course content, he demonstrates that the first kinds of courses to develop were those focusing on the "effective use of language, grammar, and style" (3), later focusing on teaching the different forms, reports, and letters routinely used in the engineering profession.

Developing later, and rapidly growing alongside the engineer writer, the surrogate engineer—the technical writer—has become at least as important in the horizons of technical communication. The conjuncture of the increased demands placed on highly specialized engineers and the growing awareness of the complexity and difficulty of encoding their ideas (meanings), gives shape to the development of technical writing as a discipline in its own right (Connors places this in the 1920s). Course work and textbooks began appearing that were directed toward the technical writing student in particular rather than toward the engineer. In spite of this specialization, the technical writer is assumed to be a mere surrogate, or stand-in, for the actual (but busy) sender, the engineer.

The technical writer's job in this period dominated by the transmission view of communication is to assure that messages are accurately encoded and that they are transmitted with minimal noise over clear channels. In fact, the professional technical writer, as surrogate engineer, is rendered essentially transparent in the process, ideally *becoming* the clear channel itself. The very definition of technical writing often affirms this commitment to the transparency of the communicator-as-channel. This is often explicit, as Michael Markel writes as recently as 1988:

> Technical writing is meant to fulfill a mission: to convey information to a particular audience or to affect that audience's attitudes in a particular way. To accomplish these goals, a document must be clear, accurate, complete, and easy to access. It must be economical and correct. The writer must be invisible. The only evidence of his or her hard work is a document that works—without the writers being there to explain. (6)

It is relatively easy to understand the location of meaning and the conception of power as they operate in this phase. Meaning is posited to be in the intentions of the sender, that is, the engineers. Meaning is simply transferred over a clear channel. Technical writers are not seen as adding or contributing to meaning. In fact, if they are, they are not doing their job! After all, they are not engineers themselves; nor are they the source of the meaning to be transmitted. Nor does meaning originate in any sense in the receiver.

Because meaning resides only in the sender's intentions, and the technical writer is merely a surrogate encoder, when communication is successful (i.e., the intended response achieved), the recognition, responsibility, and power is attributed only to the sender. However, if communication fails, it is exceedingly easy to fault the encoding process, that is, the work of the technical writer. *Miscommunication,* as this failure is called, can be attributed to the weak use of language (inadequate encoding), failure to include appropriate information (inadequate encoding), or poor standards for documentation (noisy channel) (see, e.g., Kostur and Hall).

Power, then, must be understood as possessed by the sender and measured by the ability of the message to achieve the desired result in the receiver. To communicate is to exercise power. The sender has no power if the receiver does not respond appropriately. Miscommunication, the principal measure of failure in this phase, occurs when there "is disparity between the message intended and the message received" (Kostur and Hall 19). Technical writers, who are rendered transparent and seen as contributing no meaning, possess no power (and therefore cannot exercise it) whenever communication is deemed successful. To be transparent is, after all, to provide a clear channel for the sender to exercise his or her power. Interestingly, however, if a message fails, technical writers can always be held responsible and called on to do a better job at encoding or transmission. They possess, then, a kind of negative power—by virtue of their potential status as "inadequate surrogates"—to manage the processes of encoding and transmission poorly and take the responsibility for miscommunication.

The persistence of thinking in these terms is evident in much of the professional and educational realities of technical communicators. The extent that their education focuses on stylistics, the proper use of forms, and skill at operating the technologies of communication—to the detriment of the kinds of knowledge and skills we introduce later—is testament to that persistence. Technical communicators are taught, for example, that the highest goal they can achieve is "clarity and brevity," which suggests a transparency that belies what they really do. On the job, the role of surrogate encoder is attested to by the extent that the communicator is treated as low in the organizational hierarchy, as working *for* the real sender, and as expert mainly in questions of style, form, editing, and media management. To transmit the sender's meaning as a perfectly executed message is the role of this communicator.

CHANGING CONCEPTIONS OF MEANING AND POWER: TRANSLATION

The second of the views specified at the outset, the translation view of communication, a view characterized by a fundamental concern over the constitution of meaning in messages in which power is negotiated between sender and receiver, has not been as clearly delineated as the transmission view. There are numerous contenders in the struggle to define the view developed in contradistinction to the transmission view of communication, and the successor has not yet been fully agreed on. There are in our reading several characteristics that the approaches to the second view seem to share

1. the conception of communication as a practice
2. the conception of meaning as produced through the interaction of sender and receiver
3. the conception of power as *negotiated.*

If you look back at our discussion of the transmission view of communication, you will note a conspicuous absence: The receiver in the process of communication is absent in any way other than as passive recipient of the communicated or miscommunicated message. Receivers add no meaning; they have no power. Reception is considered to be essentially unproblematic. If the message is encoded properly and sent over a clear channel, it should have the desired impact on the receiver. In contrast to this view, theorists of the translation view consider the activity of the receiver to be just as constitutive of the communication process as that of the sender. Communication is not a linear process that proceeds from sender to receiver, but a process of negotiation in which sender and receiver both contribute—from their different locations in the circuit of communication—to the construction of meaning. The nature of this process of negotiation can be understood by illustrating its operation in Stuart Hall's elaboration of what he has called a theory of "encoding and decoding."

Hall describes communication as a practice in which sender, message, and receiver are but "different moments" in a "complex structure of relations." Communication is "a structure produced and sustained through the articulation of linked but distinctive moments—production, circulation, distribution/consumption, reproduction" ("Encoding" 128). Each moment has its own distinctiveness and modality and contributes to the circulation that constitutes the communication. Hall describes it this way:

> The process . . . requires, at the production end, its material instruments—its "means"—as well as its own sets of social (production) relations—the organization and combination of practices within

> media apparatuses. But it is in the *discursive* form that the circulation of the product takes place, as well as its distribution to different audiences. Once accomplished, the discourse must then be translated—transformed, again—into social practices if the circuit is to be both completed and effective. If no "meaning" is taken, there can be no "consumption." If the meaning is not articulated in practice, it has no effect. ("Encoding" 128)

The acts of encoding and decoding are thus both active processes in the circuit of meaning production. The sender encodes meaning (meaning one) based on the frameworks of knowledge, relations of production, and technical infrastructure within which the sender operates. A *meaningful* product is produced (a technical report, for example). But the receiver also actively decodes a meaning (meaning two) based on potentially *different* frameworks of knowledge, relations of production, and technical infrastructure. There is no necessary correspondence (or symmetry) between meaning one and meaning two, because each operates semiautonomously. It is as though the practices of encoding and decoding are practices of *translation,* from social practices to discourse and then back into social practices.

When there is symmetry between the translation processes, we can talk about equivalence between the two moments—a way of rethinking the concept of *understanding.* And when there is a lack of symmetry, we can talk about a lack of equivalence—a way of rethinking the concept of *misunderstanding.* Misunderstanding cannot be explained fully by inadequate skill at encoding or by the presence of noise in the channel. Any asymmetry can also be understood as an outcome of alternative practices of encoding and decoding (Morley).

Some translation approaches continue to use a concept such as misunderstanding because they persist in privileging the encoding process. Hall, for example, posits the encoded meaning (meaning one) as the "dominant or preferred meaning" ("Encoding" 134). Then in comparing the symmetry between the preferred, encoded meaning and various decoded meanings, decodings are determined to be within the dominant, or preferred, code (dominant decoding); against it (oppositional decoding); wildly unrelated to it (aberrant decoding); or in a negotiated relationship to it (negotiated decoding) (Morley).

Some translation approaches have sought to dispense with the privileging of encoded meanings and render both moments as more *equally* constitutive. These approaches, such as that of John Fiske *(Television),* use conceptions of an "open text," conceptions such as polysemy and Bakhtin's heteroglossia. Heteroglossia asserts that "all utterances . . . are functions of a matrix of forces practically impossible to recoup" (qtd. in Fiske, *Television* 89). Polysemy asserts that a text is not merely a bearer of meanings. Rather, a text identifies and limits "an arena within which the meanings can be found. . . . [W]ithin those terms there is considerable space for the negotiation of meaning" (84). The more open a text, the greater the range within which receivers are free to make their own meanings.

Meanings are thus located in several places: in the practice of encoding, in the discursive product, and in the practice of decoding. In the passage of these forms, "no one moment can fully guarantee the next moment" (Hall, "Encoding" 129). Meaning is fluid and elusive, never really fixed at any moment.

Power is displaced and fluid along with meaning. There is power in the practice of making meaning. Because both encoders and decoders generate meaning, both exercise power. This is no longer simply the power of sender over receiver but the differential power of each to bring their own context to bear in the making of meaning (Fiske, *Television*).

Despite the fluidity of meaning, the translation view deals uneasily with differential relations of power. The receiver can work with the product (or text) only as it has been encoded, and that limits the openness of the text. This situation still privileges the practice of encoding. As Hall puts it,

> Polysemy must not, however, be confused with pluralism. Connotative codes are not equal among themselves. Any society/culture tends, with varying degrees of closure, to impose its classifications of the social and cultural and political world. These constitute a *dominant cultural order*, though it is neither univocal nor uncontested. This question of the "structure of discourses in dominance" is a crucial point. The different areas of social life appear to be mapped out into discursive domains, hierarchically organized into *dominant or preferred meanings*. ("Encoding" 134)

These dominant, or preferred, meanings must *work* to exercise power—to bring decodings into symmetry with the encodings. But decoders—always active in the decoding process—variously exercise their power to disrupt the circulation of power by decoding differently and articulating meanings differently into practice. Communication is thus *an ongoing struggle for power,* unevenly balanced toward encoding.

Currently, the field of technical communication seems to be struggling with (sometimes against) the implications for the role of the technical communicator as translator. The most obvious marker of this shift is that the technical *writer* becomes the technical *communicator* with the recognition that communicators have something to add beyond skillful encoding and clear channel. But there is much more than a name change here. To be expert in the practice of communication, to be a *communicator* in the process, signifies changes in understanding the power of the receiver as well as of the technical communicator—changes that open a virtual Pandora's box that can never again be closed.

There are a number of new things to attend to now (sometimes old things in new ways):

1. Because the process of encoding is always a process of trying to fix already slippery meanings, it is important for the communicator to understand the context of the sender. Hence familiarity with the technical field of the sender will work to ensure that in the translation process, the preferred meanings are the ones that get fixed.
2. Because the process of encoding is always an imperfect translation, it is important for the communicator to become expert at understanding and manipulating language as polysemic. Hence familiarity with the principles of rhetoric and composition and skill at using their tools will work to ensure that the communicator will know how to fix meanings.
3. Because the receivers of technical communications have the power to decode differently, depending on the contexts within which they operate, the communicator must understand how those audiences decode. Hence rhetoric (as the art of persuasion), composition, audience analysis, and reader-response research will help to ensure that communicators know how to encode such that particular audiences are most likely to decode symmetrically.
4. Further, once it is recognized that there is always a struggle to fix otherwise slippery meanings, the communicator must acknowledge and work with the differential relations of power within which sender and receiver operate. Hence attention to power and ethics is essential.

These concerns all become well represented in the field of technical communication from the 1950s on, although attention to power and ethics seems least represented, for reasons discussed later. The evidence of these changing priorities can be seen in the growing recognition of the unique contribution that can be offered by technical communicators as experts rather than as surrogates. This recognition is self-reflexive, which may account for the developing professionalization of technical communication. Evidence can also be seen in the changing textbooks and instruction in technical communication (Connors; Souther). Although stylistics, grammar, editing, and the use of media still play a major part in the education of technical communicators, it has also become essential to add to their educational repertoire work in rhetoric and composition, linguistics, problem solving, audience analysis, and ethics.

There are still employers, educators, and students whose understanding of communication is linked to the thinking of the first, or transmission, view. They have difficulty understanding the role of all this theory in just getting the job done (see, e.g., Vaughan 80). But what they fail to understand is that to execute the job with sophistication—to work toward the negotiation of symmetry between encoder and decoder—the theory must be brought to bear on the practice of communication. That requires attention to the complex and variable contexts within which senders and receivers produce meanings and how those contexts connect in the circuits of meaning and power.

Technical communication education is still in the process of sorting out those connections, establishing the balance between theory and practice. Becoming well established is the need to *go on theorizing,* to recognize that technical communication is not simply a skill but an academic and practical discipline that requires us to push the boundaries of theory if we are to understand what works and why.

But there is more to say about meaning, power, and ethics. The promise (for some, the pestilence) released from the Pandora's box of the translation view of communication is the power of the technical communicator as translator. Given the fluidity of meaning and the polysemy of any text, a translator can never be transparent. Lawrence Grossberg describes the position of the translator in this view: "Translation involves the retrieval and reconstitution of two different traditions, of two different sets of possibilities and closures. It always involves us in compromise, not only of the text's language, but of the translator's as well" ("Language" 221). The technical communicator, by virtue of the nature of the language, then, *must* add, subtract, select, and change meaning. This ushers in the recognition that the communicator, too, exercises power, that is, the communicator—operating from within a different context—makes meaning too. That recognition requires attention to ethics grounded in an understanding of how power works.

There seems to be a subtle recognition in the field that the communicator has power, but coming to terms with the nature of that power gets lost in the demarcation of encoding and decoding, of sender and audience, as the principal sites of investigation. Most educators acknowledge that it would be a good idea for students to understand politics, power, and ethics, but there is very little explanation offered to suggest what they might do with that knowledge on the job. But one thing is certain: A technical communicator cannot be just a technical writer anymore. What, then, do technical communicators offer? We think there are some answers suggested if we look ahead through the lens of ongoing theorizing in communication.

Changing Conceptions of Meaning and Power: Articulation

The third of the views specified at the outset, the articulation view of communication, a view characterized by concerns with the struggle to articulate and rearticulate meaning and relations of power, can be delineated in contrast to both the transmission and the translation views. The transmission view acknowledges that senders do have meanings that they desire to encode and that they do often desire a particular response to that message from the receiver. However, the transmission view limits our recognition of the full fluidity of meaning. The translation view reconstitutes transmission to add an understanding of the receiver's contribution to the constitution of meaning and introduces the constitutive role of a mediator. However, translation based on the model of encoding and decoding limits our understanding of the full authorial contribution and power of the mediator.

The translation view opens the space for the attribution of authorial power (the Pandora's box) but leaves it undertheorized. The opening is evident in Grossberg's assertion (cited previously) that the language of the translator must be taken into consideration. The way through that opening is provided in the very language of encoding and decoding, specifically in thinking through Hall's suggestion that meaning is "articulated in practice" and that meaning and discourse are "transformed... into social practices... if the circuit [of meaning] is to be both completed and effective" ("Encoding" 128). The articulation view allows us to move beyond a conception of communication as the polar contributions of sender and receiver to a conception of an ongoing process of articulation constituted in (and constituting) the relations of meaning and power operating in the entire context within which messages move. That context includes not just the context of the sender and receiver (the frameworks of knowledge, relations of production, and technical infrastructure) but of the mediator(s) as well. And *mediator* here can no longer be thought of as just the technical communicator but as the channels (including media and technologies) of transmission as well.

Articulation is a concept that has been drawn from the work of Antonio Gramsci, considered by Emesto Laclau, influenced by structuralism (especially Althusser) and postmodemism (see, e.g., Deleuze and Guattari), and developed into an identifiable theoretical position by Hall ("On Postmodemism"; "Race"; "Signification"). Grossberg has elaborated on the role of power in this position ("Critical Theory"). Articulation asserts that any identity in the social formation must be understood as the nonnecessary connection between the elements that constitute it. Each identity is actually a particular connection of elements that, like a string of connotations, works to forge an identity that can and does change (Hall, "Signification"). An identity might be a subject, a social practice, an ideological position, a discursive statement, or a social group. The elements that constitute these identities are themselves identities; therefore, they too must be understood as nonnecessary, changing connections between other elements. The way in which elements connect or combine is described as an articulation. As Jennifer Daryl Slack has described, articulations, the connections between elements that forge identities, have the following characteristics:

> (a) Connections among the elements are specific, particular, and nonnecessary—they are forged and broken in particular concrete circumstances; (b) articulations vary in their tenacity; (c) articulations vary in their relative power within different social configurations; and (d) different articulations empower different possibilities and practices. (331)

Any identity might be compared to a train, which is constituted of many different types of train cars in a particular arrangement (or articulation). Each car is connected (or articulated) to another in a specific way that, taken as a whole (as a series of articulations), constitutes the identity *train.* Any specific train is thus a specific, particular set of articulations—an identifiable object with relatively clear-cut boundaries. But these specific articulations are nonnecessary; that is, there is no absolute necessity that they be connected in just that way and no guarantees that they will remain connected that way. So, for example, we could disconnect (disarticulate) and reconnect (rearticulate) cars in a different order to constitute a new identity *train.*

To say that articulations vary in their tenacity is to acknowledge that some connections are more difficult to disarticulate/rearticulate than others. Yard police, for example, may or may not let us in to change the order of the cars. Or the kinds of connections between the cars may be variously difficult to manipulate.

Some articulations are more resistant to rearticulation than others; that is, some are more *tenacious* than others. When a connection between elements is particularly resistant, the identity *train* remains intact and effective over a long period. When an articulation is effective, it is said to be powerful in that it delineates what is real and possible from what is not. Different arrangements make possible different possibilities and practices. If we disarticulate the engine, for example, the rest of the train will not move. And, in the process, we may have rearticulated the elements in such a way as to necessitate a new identity. Is a string of cars without an engine a train? Is a single engine a train? We take the answers to both to be maybe. On the other hand, a disarticulated car of the type that usually completes a train will probably not be thought of as constituting the identity *train.* The term *caboose* might have to suffice. But a train without a caboose is usually still thought to be a train.

Articulation thus points to the fact that any identity is culturally agreed on or, more accurately, struggled over in ongoing processes of disarticulation and rearticulation. For example, clearly, one element of what makes a train a train (and not, say, just a caboose) depends on our agreed-on cultural conception of *train.* To stretch this a bit, we could say that we have an ideology regarding what we empower as a train. The ideology of *train* articulates to the arrangement of the cars such that we may call a lone engine a train but not a lone caboose. But that ideology is itself an identity constituted by its articulations, one of which is the past practices of putting trains together. Given changes in those practices, say, for example, giving cabooses their own little engines to get around, we may rearticulate our ideology of *train* such that lone cabooses are more like lone engines and deserve, perhaps, the status, *train.* Alternatively, we may alter the identity *train* by working to rearticulate it on ideological grounds alone. We may, as teachers, for example, decide to teach people a different definition (identity) of *train* so that a lone engine or a lone caboose is rearticulated as constituting the identity. The success of our attempts at rearticulating identities, whether purposeful or not, depends on the tenacity of the various articulations that constitute it at any particular conjuncture.

To extend this now beyond more easily identifiable identities, social practices, ideological positions, discursive statements, social groups, and so on are also articulated identities whose meanings are continually and variously rearticulated. Dictionaries define the most widely accepted (or acceptable) identities, but there are frequently different, alternative articulations that are either archaic or emerging. One need only read a bit of the *OED* to begin to get a feel for how dramatically articulations can change (although the *OED* only hints at the range of connections that constitute the articulations). Raymond Williams's *Keywords* tracks changing articulations of key identities in Western thought and provides excellent cases of rearticulation.

The concepts of meaning and power are dramatically refigured in articulation theory. Meanings cannot be entities neatly wrapped up and transmitted from sender to receiver, nor can they be two separate moments (meaning one contributed by the sender and meaning one contributed by the receiver) abstractly negotiated in some sort of a circuit. Like any identity, meaning—both instances and the general concept—can be understood as an articulation that moves through ongoing processes of rearticulation. From sender through channels and receivers, each individual, each technology, each medium *contributes* in the ongoing process of articulating and rearticulating meaning. Power is no longer understood as simply the power of a sender over a receiver or as the negotiated symmetry of the sender's or receiver's meanings but as that which draws and redraws the lines of articulation. As Grossberg has put it, power "organizes the multiplicity of concrete practices and effects into predefined identities, unities, hierarchical categories, and apparently necessary relationships" ("Critical" 92). Power is thus what works to *fix meanings,* that which empowers some possibilities and disempowers others. Grossberg explains that empowerment is "the enablement of particular practices, that is, as the conditions of possibility that enable a particular practice or statement to exist in a specific social context and that enable people to live their lives in different ways" (95).

We can expand our understanding of the role of the technical communicator and of technical discourse significantly by tracking the implications of an articulation view of communication. First, by using the lens of articulation theory, we have here been able to track the changes in the theory and practice of technical communication as themselves rearticulations of elements (or identities) such as technical communicator, meaning, author, channel, sender, power, receiver, and so on. Second, however, that very lens works to rearticulate the location of the technical communicator in the process of communication, specifically in that technical communicators must now be understood as articulating and rearticulating meaning in (and variously contributing to or changing) relations of power. To gain access to those rearticulations, we will again consider the question of authorship as raised by Foucault at the beginning of this article.

It is tempting here to begin to lay out all of the elements that articulate to the notion of *author* as it moves through the stages of transmission, translation, and finally to articulation itself. These articulations would include elements such as the conception of authors as individuals, individuals as the source of meanings, the conception of meaning as a fixed entity—the practice of attributing ownership to ideas, capitalist relations of property and appropriation, a notion of the power of ideas, and a particular conception of progress ("if it's new, it's better"). However interesting that task might be, we must limit our treatment here to some very specific articulations that direct our attention to the questions of meaning and power in the theory and practice of technical communicators.

In the transmission view of technical communication, authority is articulated to scientific and technical discourse as an objective and neutral reporting of facts. Humanities types may author meaning, but scientists, engineers, and, by extension, technical writers, merely (albeit skillfully) represent what is already objectively "out there." These are not meanings, but objective, disembodied facts. Consequently, technical communications (like the posters or contracts mentioned by Foucault) often have no authors. When technical documents are *conveyed* by named individuals, these are again not authors in the sense of originating meaning—these are simply not discourses that produce an author. Even in these cases, however, for reasons considered later, technical communicators are rarely listed among the conveyers.

Technical documents and writing in science and engineering do often name authors (what Foucault calls the writer). In this case, the author remains the sender in the transmission sense but, articulated now to the concept of conveyance of scientific fact, as an authority. Rarely, again, is authorship in these cases extended to the professional technical communicator. In part, the attribution of authorship here to the scientist or engineer at the expense of the technical communicator must be explained in terms of the tenacity of other articulated elements: the neutrality of scientific discourse, the practice of attributing ownership to ideas, a conception of invention as the expression of individual genius, capitalist relations of property and appropriation, and the persistence of the elevation of the scientific discourse over humanistic discourse (see Horkheimer and Adorno). In other words, specific relations of power articulated to a particular conception of science account for the specific identity of authorship in the sciences and the exclusion of the technical communicator from that attribution.

To evoke *author* in theory or practice from within the transmission view evokes, like a chain reaction of connotations, all these articulations, which struggle—whether purposefully or not—to hide the work that goes into fixing the identity of that work. These articulations are nonnecessary; that is, there is no necessity that they be connected in just this way and no guarantee that they will remain connected in this way. Indeed, translation works to rearticulate the question of authorship, although its challenge is incomplete.

Although the translation view suggests a more elevated role for the translator, it does not grant authority. To put it another way, the translator is seen as an expert, but only in mediating, not authoring, meanings. This is even the case in the humanities, where debates ensue over whether or not to give translators the same credit in tenure and promotion reviews as authors. In technical communication, the unique skill of individuals may be recognized as acts of mediation, but as an activity, the discourse still does not grant them authorship. Again, we suggest that this is in part due to the tenacity of some of those same relations of power discussed earlier: the practice of attributing ownership to ideas, the conception of invention as the expression of individual genius, and capitalist relations of property and appropriation.

By resting on the conception of author as articulated to the contribution of meaning, by challenging the articulation to differential relations of power between sender, translator, and receiver as being somehow evident, and to the conception of science as objective fact finding, we would advance the rearticulation of technical communicators (along with media and technology) as having authorial power. We cannot grant technical communicators status as authors merely in the scientific sense of *conveyers of fact.* That would be to deny the insight of even the translation view that asserts that the discourse of the translator (whether the translator be scientist, technology, medium, or technical communicator) must be understood as involved in the compromise. Rather, technical communicators are theoretically situated in the process of articulating meaning just as prominently as are the sender and the receiver. The process of communication is then not simply a transmission or a translation but an articulation of voices, much like what Bakhtin has characterized as the orchestration of "heteroglot, multi-voiced, multi-styled, and often multi-languaged elements" (265).

It should be obvious that different articulations empower different possibilities and disempower others. When technical communicators are not articulated to authorship, their possible contributions are severely constricted. Whether they desire it or not, technical communicators are seen as variously adding, deleting, changing, and selecting meaning. Again, whether they desire it or not, they are always implicated in relations of power. Their work is at least *complicit* in the production, reproduction, or subversion of relations of power. This is necessarily the case, even when the acceptance of the transmission or translation view may occlude the nature of the work that they do. Technical communicators *are* authors, even when they comply with the rules of discourse that deny them that recognition. When they are denied that recognition, the measure of their success can only be complete compliance with the articulations of meaning, power, and authorship from the standpoint of the transmission and translation views.

The consequences of extending authorship to technical communicators are significant. With the recognition that the communicator articulates and rearticulates meaning comes the responsibility for that rearticulation. No contribution is really transparent; it is only rendered transparent in relations of power. So, just as the power of technical communicators is recognized (as they are empowered), so too must they be held responsible.

IMPLICATIONS FOR PEDAGOGY AND PRACTICE

We heard recently of an industry recruiter who—venting some frustration over graduates knowing more theory than was good for them on the job—said, "We want robots!" This frustration has, we submit, several sources. First, and most obvious, we take this to be a plea for technical communicators to perform their transmission function well. We would not dispute the need to be able to perform skillfully using effective grammar, editing, media management, and so on.

But there is more in the recruiter's frustration. Second, then, this plea points to the fact that the field is growing rapidly in the tension between transmission, translation, and articulation. Although that tension is generative, it does not result in easily written job descriptions, clear definitions of the technical communicator's role, task, and ethos. Sometimes there is a lack of clear vision and agreement—among practitioners and their employers—about what is expected of a technical communicator and what it is he or she has to offer. In addition, however, that plea for robots suggests that there exists a particularly tenacious articulation between the conception of communication as the transparent transmission of messages, the neutrality of science and engineering, and perhaps even of the ethical neutrality of the ethics of capitalism. In fact, to behave as such a robot is to be complicit with the meanings thus articulated.

It is possible to look at some of the turmoil in the education of technical communicators and some confusions about the work of these graduates in terms of a field trying to come to terms with consequences of the technical communicator as author. The difficulties are twofold: On the one hand, the theoretical development of an articulation view has not advanced far enough to form a firm foundation for pedagogy and models of work. On the other hand, the changes that would result from this rearticulation—although theoretically and practically defensible—are not likely to come easily.

Nevertheless, because professional communicators contribute to the process of articulating meaning, whether they choose to or not, they must be able to analyze critically the ethical implications of the meanings they contribute to. Such knowledge is all the more important given the current tendency to define their work as (ethically) transparent. In a sense, technical communicators need to be shaken from the somnambulistic faith that their work is ethically neutral. Steven Katz's examination of a virtually "perfect" technical document proposing changes in a vehicle designed to asphyxiate prisoners during the Nazi holocaust ought to put an end to any assertion of ethical neutrality. It is not simply *how well* we communicate that matters. *Who* we work for and *what* we communicate matters.

The nearly ubiquitous calls for technical communicators to learn more about the technical content of their work (see, e.g., Institute of Electrical and Electronics Engineers), even to participate in the early stages of project design, can be understood as easily articulated to the conception of the communicator as author. Such technical knowledge can provide the backdrop for sound, ethical decision making, as well as for competent transmission and translation.

In addition to ethics and technical knowledge, it seems equally essential that technical communicators have a superior grasp of the relationship between technology and discourse and between science and rhetoric (Horkheimer and Adorno; Miller; Wells; Sullivan; Katz). It is essential that we learn to analyze critically the articulations evoked in the language of technology and science. In a sense, technical communicators need to be shaken from the somnambulistic faith that their work is linguistically neutral.

Finally, we would add to the education of technical communicators knowledge of how organizations operate—in the form of organizational communication or organizational behavior. It is remarkable how little most of us understand the relationship between power, knowledge, and organizations. It is time that we give up the faith that the goal of communication is always clarity and brevity. In practice, the politics of organizations and organizational politics often have as their goals limiting, obscuring, or hiding information (Wells; Katz; Butenhoff). Naïveté about how organizations work articulates well to the myth of the technical communicator as engaging in an ethically and linguistically neutral activity.

To send out technical communicators with this kind of knowledge is to send them out armed.[1] It is impossible for technical communicators to take full responsibility for their work until they understand their role from an articulation view. Likewise, it is impossible to recognize the real power of technical discourse without understanding its role in the articulation and rearticulation of meaning and power. This understanding would thus empower the discourse of technical communicators by recognizing their full authorial role.

NOTE

1. We invite our readers to explore the consequences of this view for the role, task, and ethos of technical communicators as advocates for their constituencies: their employers, clients, and audiences. As advocates, they would be more like lawyers than their current status acknowledges. Although technical communicators have less in terms of codified law or precedent on which to draw, they could be understood as advocating for, counseling, advising, defending, or building cases. This change in status complicates the relationship to their constituents: The counsel of communicators might be accepted, rejected, or resisted (or litigated against!). But just as a lawyer's duty is to inform employers or clients of the possible consequences of their actions, so too should it be the technical communicator's duty to inform employers or clients of the consequences of their rhetoric!

In addition, this view suggests that the expertise of technical communicators is applicable to the articulation of meaning well beyond the confines of science and engineering (or business). Instead, its scope can easily be understood as encompassing situations in which the transmission, translation, and articulation of specialized knowledge is at issue.

Finally, we do not offer this invitation with any pretense that advocacy or authorship will *simplify* the role, task, and ethos of technical communicators. We offer no apology, however, for we are advocating here changes that are already underway, even if they are not very well understood.

WORKS CITED

Althusser, Louis. *For Marx.* Trans. Ben Brewster. New York: Random House, 1970.

Bakhtin, M. M. The *Dialogic Imagination: Four Essays.* Trans. Caryl Emerson and Michael Holquist. Ed. Michael Holquist. Austin: U of Texas P, 1981.

Butenhoff, Carla. "Bad Writing Can Be Good Business." *Readings in Business Communication.* Ed. Robert D. Gieselman. Champaign, IL: Stipes, 1986. 128–131.

Carey, James. *Communication as Culture: Essays on Media and Society.* Boston: Unwin Hyman, 1989.

Connors, Robert I. "The Rise of Technical Writing Instruction in America." *Journal of Technical Writing and Communication* 12 (1982): 329–352.

Cooper, Marilyn M. "Model(s) for Educating Professional Communicators." *The Council for Programs in Technical and Scientific Communication [CPTSC] Proceedings* 1990. San Diego, CA, 12 October. Ed. James P. Zappen and Susan Katz. CPTSC, 1990. 3–13.

Deleuze, Gilles, and Felix Guattari. "Rhizome." *Ideology and Consciousness* 8 (1981): 49–71. "Ferment in the Field" [Special issue]. *Journal of Communication* 33.3 (1983).

Fiske, John. *Introduction to Communication Studies.* New York: Methuen, 1982.

———. *Television Culture.* New York: Methuen, 1987.

Foucault, Michel. "What Is an Author?" *Language, Counter-Memory, Practice: Selected Essays and Interviews.* Ed. Donald F. Bouchard. Ithaca, NY: Cornell UP, 1977. 113–138.

Gramsci, Antonio. *Selections from the Prison Notebooks.* Ed. and trans. Quentin Hoare and Geoffrey Smith. London: Lawrence & Wishart, 1971.

Grossberg, Lawrence. "Critical Theory and the Politics of Empirical Research." *Mass Communication Review Yearbook.* Ed. M. Gurevitch and M. R. Levy. London: SAGE, 1987. 86-106.

———. "Language and Theorizing in the Human Sciences." *Studies in Symbolic Interaction* 2 (1979): 189–231.

Hall, Stuart. "Encoding/Decoding." *Culture, Media, Language.* Ed. Stuart Hall et al. London: Hutchinson, 1980. 128–138.

———. "On Postmodemism and Articulation: An Interview with Stuart Hall." *Journal of Communication Inquiry* 10 (1986): 45–60.

———. "Race, Articulation and Societies Structured in Dominance." *Sociological Theories: Race and Colonialism.* Ed. UNESCO. Paris: UNESCO, 1980. 305–345.

———. "Signification, Representation, Ideology: Althusser and the Post-Structuralist Debates." *Critical Studies in Mass Communication* 2 (1985): 91–114.

Horkheimer, Max, and Theodore Adomo. *Dialectic of Enlightenment.* New York: Herder & Herder, 1972.

Institute of Electrical and Electronics Engineers (IEEE). *The Engineered Communication: Designs for Continued Improvement.* Vols. 1 and 2. International

Professional Communication Conference Proceedings. Orlando, FL, 30 Oct.–1 Nov. New York: IEEE, 1991.

Katz, Steven B. "The Ethic of Expediency: Classical Rhetoric, Technology, and the Holocaust." *College English* 54.3 (1992): 255–275.

Kostur, Pamela, and Kelly Hall. "Avoiding Miscommunication: How to Analyze and Edit for Meaning." *The Engineered Communication: Designs for Continued Improvement.* Vol. 1. International Professional Communication Conference Proceedings. Orlando, FL, 30 Oct.–1 Nov. New York: Institute of Electrical and Electronics Engineers, 1991. 18–25.

Laclau, Emesto. *Politics and Ideology in Marxist Theory.* London: Verso, 1977.

Markel, Michael H. *Technical Writing Essentials.* New York: St. Martin's, 1988.

McQuail, Denis, and Sven Windahl. *Communication Models for the Study of Mass Communications.* New York: Longman, 1981.

Miller, Carolyn R. "Technology as a Form of Consciousness: A Study of Contemporary Ethos." *Central States Speech Journal* 29 (1978): 228–236.

Morley, David. *The "Nationwide" Audience.* London: BFI, 1980.

Oxford English Dictionary. 1971.

Shannon, C., and W. Weaver. *The Mathematical Theory of Communication.* Champaign: U of Illinois P, 1949.

Slack, Jennifer Daryl. "Contextualizing Technology." *Rethinking Communication.* Paradigm Exemplars 2. Ed. Brenda Dervin et al. Newbury Park, CA: SAGE, 1989. 329–345.

Souther, James W. "Teaching Technical Writing: A Retrospective Appraisal." *Technical Writing Theory and Practice.* Ed. Bertie E. Fearing and W. Keats Sparrow. New York: Modern Language Association, 1989. 2–13.

Sullivan, Dale L. "Political-Ethical Implications of Defining Technical Communication as a Practice." *Journal of Advanced Composition* 10.2 (1990): 375–386.

Vaughan, David K. "The Engineer: Neglected Target of Technical Writing Instruction." *The Engineered Communication: Designs for Continued Improvement.* Vol. 1. International Professional Communication Conference Proceedings. Orlando, FL, 30 Oct.–1 Nov. New York: Institute of Electrical and Electronics Engineers, 1991. 80–83.

Wells, Susan. "Jürgen Habermas, Communicative Competence, and the Teaching of Technical Discourse." *Theory in the Classroom.* Ed. Cary Nelson. Urbana: U of Illinois P, 1986. 245–269.

Williams, Raymond. *Keywords.* London: Fontana, 1976.

Chapter Two

Extreme Usability and Technical Communication

Bradley Dilger

In the past twenty years, usability has become a more important part of the development of technological systems and technical communication, as evident from its increased role in the design of computers and other electronics, or recent articles in *Technical Communication Quarterly*. For the most part, this growth has been positive, one of the reasons for the current strength of user-centered methods of developing technology and communication. However, as usability has matured, the rise of a particular kind of usability, which I call "extreme usability," has displaced original, more complex formulations. Practitioners of extreme usability repeatedly invoke ease and "making it easy" in their definitions of usability. This is no accident: extreme usability is, in fact, usability made easy, a simplified usability profoundly and problematically distinct from the robust, more carefully developed concepts of usability from which it was derived.

Bernadette Longo has argued that a cultural studies approach to studying technical communication can make writing visible for study, demonstrating how theories and practices that shape writing are articulated. For extreme usability, this approach is especially important, since extreme usability limits the visibility of technology and technological institutions, encouraging the belief that technology is autonomous, practically or empirically beyond our capability and control. Studying the cultural and historical context of extreme usability illuminates its transfer of the transactional, consumerist logic of ease to usability, and the resultant weakening of end-user agency. Like Longo and other cultural studies theorists, I believe that technical communication scholarship must consider these issues—after all, as Robert Johnson points out in *User-Centered Technology*, the desire to address the inequitable distributions of power built into most technology and communication often motivates the user-centered methodologies usability was originally intended to support. In contrast, extreme concepts of usability facilitate a "best practices" approach to usability, which reduces user engagement, forbids considering the wider scope of culture, and limits the ends of usability to achievement of expediency.

Therefore, in this essay, I investigate the bond of ease and extreme usability from a cultural studies perspective, considering the ways the ideology of ease affects definitions of usability, methodologies for measuring usability, and the many kinds of readers and writers involved in these processes. I begin by explaining how extreme usability is derived from more robust concepts of usability. I outline the history of ease, focusing on its growth into a powerful ideology that shapes Americans' understanding of technologies, from writing to consumer goods to computers. Calling on contemporary textbooks, web design books, and other sources, I then demonstrate how extreme usability—as a supposedly measurable quality and as a methodology—reproduces the ideology of ease, consistently undermining user-centered approaches to usability. I conclude by focusing on the role of extreme usability in technical communication, arguing that by mobilizing fully developed concepts of usability, technical communication educators can prevent the considerable problems that can occur when usability is reduced to extreme usability.

THE DEFINITION OF USABILITY AND EXTREME USABILITY

Extreme usability has arisen in part as a response to the complexity of influential definitions of usability. For example, in *Usability Engineering,* Jakob Nielsen defines usability as five interconnected parts: (1) learnability, or being "easy to learn"; (2) efficiency of use; (3) memorability, or "an interface that is easy to remember"; (4) few and noncatastrophic errors; and (5) subjective satisfaction, or pleasure in use (26–33). Additionally, variations in definition persist despite repeated calls for consensus and stability, such as the codification of usability in ISO standard 9241 (Dillon 14). The ends of usability differ, too; the concept can describe a desired result of the development of technological systems and practices, a method for shaping user interfaces for machinery and software, or a philosophy for evaluating user needs (Quisenbery 82). Finally, high-tech fields like computer programming, information architecture, interface design, and consumer electronics continue to experience rapid growth, continually moving the target of usability. Extreme usability confronts this complexity and variation of definition, multiplicity of purpose, and rapid change by creating a more easily attained concept of usability.[1]

Extreme implementations of usability often streamline its complexity by using a given definition as a point of departure, but privileging one component over the others. For example, the concept of "user-friendly" design, often considered to be analogous to usability, weighs user satisfaction and error recovery heavily (*Apple Human Interface Guidelines*). Barbara Mirel argues that this pattern is common not only in theory, but in practice, and calls for the more integrated approach of concepts of usability like Nielsen's:

> Usability [. . .] involves multiple dimensions [. . .]. All are intricately intertwined. None is independent of the others. Yet, in many development contexts, a comprehensive vision of interrelated usability dimensions gets broken apart. Each dimension—ease and efficiency of use, learnability, enjoyment, and usefulness—becomes a separate objective. Design and development teams choose to build for some dimensions while neglecting others based on project deadlines, resources, and other constraints. (168)

The simplifications Mirel criticizes likely provide a more measurable, teachable, attainable usability. But a concept of usability built on one part of a multivalent definition cannot make best use of various accompanying methodologies (such as rapid prototyping, focus groups, heuristic evaluation, or formal usability testing), which together rely on the more complete theoretical foundation. And as Mirel shows, these condensations of usability are often mere expedients applied to move projects forward, not approaches engaged because of past effectiveness or demonstrated benefits.

Whitney Quisenbery argues that much of the oversimplification of usability focuses on privileging ease of use and ease of learning, concepts quite prominent in many definitions of usability (88–89). Focus on ease is, in a nutshell, the heart of extreme usability. Ease figures heavily in Nielsen's five-part definition, as well as the definition of usability with which Deborah Mayhew introduces the textbook *The Usability Engineering Lifecycle*:

> **Usability** is a measurable characteristic of a product user interface that is present to a greater or lesser degree. One broad dimension of usability is how *easy to learn* the user interface is for novice and casual users. Another is how *easy to use* (efficient, flexible, powerful) the user interface is for frequent and proficient users, after they have mastered the initial learning of the interface. (1, emphasis in original)

Mayhew's definition focuses on the role of ease in surface-level features (user interfaces). Similarly, Karen Donoghue sees extreme usability as "an experience that *makes it easy* to complete goals and accomplish tasks with a minimum of friction" (5, my emphasis). For Informatica's PowerAnalyzer, extreme usability is achieved with a single "zero-client" interface that enables "minimal training and rapid, mass deployment," using natural language and "familiar paradigms." In these and many other contexts, the concept of usability is reduced to various forms of making technologies and tasks easy, and achieving usability itself is likewise portrayed as easy. Though the examples I provide here focus on computing (user interfaces), ease has been associated with technology since the seventeenth century, when a strong bond between ease and writing was first forged. I would like to review the history of ease in American consumer culture, especially the development of ease into a powerful ideology, to show how the emergence of extreme usability continues the ease-technology connection.

Historicizing Ease, Usability, and Extreme Usability

The appearance of usability is traceable to the disciplines of ergonomics and human factors engineering, which emerged during and immediately after World War II, in response to the need to make complex military equipment less difficult to operate (Barnum xiii). But a concept of usability in which ease is foundational has roots in consumer culture. In the first part of the twentieth century, marketing of new mass-produced consumer products promised to deliver ease by reducing the amount of labor necessary to complete a task. However, consumer products were not easy to operate by current standards, and were rarely marketed as such. For example, while early electric washing machines made laundry less time-consuming and less back-breaking than using the washboard, tub, and wringer, their operation required manipulation of a sophisticated set of controls. Washing, rinsing, and wringing out clothes sometimes required reconfiguring equipment with tools or complicated pulleys and levers—and exposed machinery often caused horrific injuries (Maxwell). While there were exceptions, like Kodak cameras, marketed with the slogan "You push the button, we do the rest," for the most part, the attitude toward contemporary technological products was highly pragmatic. As long as products enabled work, increased the amount of work possible in a given time, or made work less demanding, they were acceptable, even with massive difficulty and complexity. This attitude was reflected in the work of contemporary writers like Christine Frederick, whose *Household Engineering* proposes adopting Taylorist methodologies to the home, and prefigured the language of usability by calling for better designed home workspaces that enabled "*comfort* in use." Importantly, Frederick was also a huge proponent of consumption: she encouraged American homemakers to alleviate their overwhelming amount and difficulty of housework by purchasing new machines—bread mixers, vacuum cleaners, fireless cookers, and electric mixers.

Consumer products gained what might be called "usability" only gradually, as the cultural power of ease grew and its role in product design and marketing increased. While lessons learned in the new disciplines of human factors undoubtedly had some effect in civilian spheres, consumer products changed on their own as well. Beginning a process that would span the next fifty years, from about 1925 to 1975, direct manipulation or electric controls—like the Edsel's push-button transmission—gradually replaced arrays of levers, valves, and switches. As more automatic or "computerized" products appeared, marketing emphasized not only labor-saving properties, but ease of use—simple operation, convenience, and speed—even the "life of ease" represented by the American dream. With an increasing number of technological devices in both home and workplace environments, and more appearing often, thanks to the endless march of progress, the complexity and difficulty of any given machine had to be minimized to ensure consumer goods could deliver the ease which they promised.

But this nascent proto-usability was limited, often along gender lines: For women and children at home (and men outside of work settings), making things easy to use was normal, a matter of foregrounding the simplicity of the already simple. For men at work, technological sophistication retained complexity and difficulty, and early proto-usability efforts (in "human factors" or "ergonomics") sought ease of use only in the service of increased functionality and productivity—as in Frederick's view of household engineering. Regardless, demands for ease grew, and it became an ideological power which influenced the development of numerous household goods, and bolstered the explosion of convenience products and services.

The introduction of graphically oriented desktop computing—popularized by the 1984 introduction of the Apple Macintosh operating system—extended demands for ease and usability to nearly all kinds of technological products in all kinds of situations. Like many proponents of usability, Donald Norman contended that making technology easy to use increased productivity by eliminating errors and accelerating learning. His 1987 *The Psychology of Everyday Things,* a watershed for ease and usability, includes numerous statistics that show the labor costs recovered by shaving a few seconds off frequently repeated tasks. But resistance to products that called attention to their easy nature, like the Mac OS, was notable; opponents of usability considered its pursuit wasteful, and questioned the integrity and effectiveness of technological systems that required it. John C. Dvorak invokes gender roles in his attack, calling the Macintosh "effeminate" and championing its competitor, the IBM Personal Computer (PC), as "a man's computer designed by men for men" (cited in Gelernter 40). For Dvorak and others, serious work didn't need usability and instead kept its focus on the bottom line. But the opponents of usability were unable to overcome the massive power of ease, and it has become the most dominant force shaping the design and use of technological systems—following the transactional logic of consumer culture in which ease first developed.

Extreme usability is the latest manifestation of ease, providing a conduit for extending its simplicity, comfort, expediency, and pragmatic character to increasingly sophisticated technological systems and devices. However, extreme usability also extends the ideological framework of ease as well, bringing the assumption of a commercial context, lack of critical engagement, and desire for speed and convenience typical of consumer culture to our understanding of technology. Like ease, extreme usability encourages out-of-pocket rejection of difficulty and complexity, displaces agency and control to external experts, and represses critique and critical use of technology in the name of productivity and efficiency. Usability is, by definition, heavily pragmatic, but complex concepts like Nielsen's moderate this pragmatism through interdependence with other qualities (such as user satisfaction), and by making usability one part of a larger process of user-centered design and development. In contrast, extreme usability's monolithic embrace of ease lacks similar checks and balances and is less suitable for the complexity of technological systems—including reading and writing—that map poorly onto the transactional logic of "making it easy."

The etymology of "extreme" shows the transfer of the paradoxical ideology of ease to usability. Recently emerged meanings of "extreme" connote technological advancement, atypical power, and a desire for risk and innovation that makes "extreme" sports, technologies, or products unsuitable for use by individuals uncomfortable with risk or innovation.[2] On the one hand, it seems inconsistent to apply this sense of "extreme" to usability—why would sophisticated, cutting-edge technology be extremely usable? But as was the case with consumer goods and technologies that are made easy, the assumption is that extreme usability shifts the danger of extremity to the expert designer, writer, or software programmer, making the power and performance of "extreme" technology accessible to the novice user. The ideology of ease assures the novice user that the concomitant loss in power and agency is insignificant, and any increased cost sustainable as well. This disengagement all but eliminates the possibility of a technology centered around user needs, undermining a robust methodology of usability.

Extreme usability appears both as a quality of technological systems or consumer goods and as an approach to attaining and evaluating usability that applies a restricted definition of usability to simplify, reduce the cost of, and expedite the process of usability assessment and implementation. I would like to discuss both forms in depth.

EXTREME USABILITY AS A QUALITY POSSESSED BY TECHNOLOGY

Jakob Nielsen has done more to popularize usability than anyone, and his widely read *Designing Web Usability* has arguably provided the most widely known theories of usability. Somewhat surprisingly, it makes only passing mention of Nielsen's large body of sophisticated and nuanced research into usability.[3] Though the book does not use the phrase, I believe that *Designing Web Usability* advances extreme usability, since its dominant concept of usability is generally congruent with the properties of ease, as it developed in consumer culture. Indeed, ease has a high profile in the text, appearing primarily as "the practice of simplicity," which is not only the subtitle of the book but an excellent summary of its ethos. As with ease, simplicity repeatedly beats complexity for a variety of reasons, not the least of which is achieving speed (22, 42–51). *Designing Web Usability* also speaks out against the complexities of metaphor in writing or site design: "Users don't live in the metaphor world, they live in the real world. [. . .] it is usually better to be very literal" (180). For web-based text, Nielsen's work argues that usability is achieved through brevity, recommending "no more than 50 percent of the text you would have used to cover the same material in a print publication" (101). Writers should also cultivate "scannability" by highlighting keywords using bold formatting, bullet points, and writing informative, literal headlines and subheads (104–106). Paragraphs should contain one idea, using topic sentences and simple sentence structures (111). *Designing Web Usability* advocates a journalistic writing style, including preference for the inverted pyramid (112), but also reminiscent of concepts of "good writing" codified in American composition classrooms during the nineteenth century—which, as I have argued elsewhere, were shaped predominantly by the concept of ease.[4]

Most problematically, in *Designing Web Usability* Nielsen presents the Web—and usability—as if both were limited to no-nonsense facilitation of electronic commerce: "Usability rules the Web. Simply stated, if the customer can't find a product, then he or she will not buy it. [. . .] While I acknowledge that there is a need for art, fun, and a general good time on the Web, I believe that the main goal of most web projects should be to make it easy for customers to perform useful tasks" (9, 11). In recent interviews and "Alertbox" columns on his website Useit.com, Nielsen has softened this hyperpragmatic stance, allowing for the diversity of activities that occur on the Web ("Interview," "User Empowerment"). However, he continues to assume a commercial context. One "Alertbox" focusing on improving home pages offers first, "Emphasize what your site offers that's of value to users and how your services differ from those of key competitors" ("Ten Most Violated"). This focus on commerce reduces the effectiveness with which usability can be considered as a dimension of technical communication, which is obviously not necessarily transaction-oriented; web pages designed for individuals, courses, or other purposes are difficult to evaluate using these criteria. Overall, *Designing Web Usability* advances a concept of usability similar to Donoghue's "experience that makes it easy to complete goals and accomplish tasks with a minimum of friction" (5). In Nielsen's vision of the Web, the ideology of ease is king, and both reading and writing mirror the logic of consumer culture.

Donoghue, who uses the phrase "extreme usability" several times, focuses on the larger framework of user experience, defined as "the behaviors and attitudes of end users and their incentives to actually use the system" (xviii). For Donoghue, ease of use is the most important attitude associated with user experience and her concept of usability:

> Ease of use should be embedded in the DNA of the user experience and intrinsic to its development process. It should be present in every atomic action in the experience, in each and every click. [. . .] Disney World is an environment where entertainment and suspension of disbelief are maintained through the entire experience. Drop a paper cup on the ground and a smiling Disney character walks by and—in a single graceful gesture designed to maintain the suspension of disbelief—the cup is swept away. User experiences should strive to be this proactive, transparent, and useful. (50)

Evaluating Disney World as the ultimate user experience shows the extremity of Donoghue's concept of usability and user experience. The example suggests a tremendous cost for achieving usability: consider the amount of effort needed to maintain the environment she portrays. Usability and user-experience design seem to be a dirty job—cleaning up after other people who can't be bothered to clean up after themselves, while appearing happy despite the menial nature of the work. What matters is the ease of use apparent to the clients or the users of a system; what's behind the mask of the interface (or in this case, the Disney character) is not at issue. In patterns similar to those Evan Watkins observed in *Throwaways,* the cost of usability is downplayed, and a lot of labor is made invisible.

Facilitating fun and enjoyment, less emphasized in *Designing Web Usability,* is critical for Donoghue's idea of usability. Once again, a single dimension of usability present in the complex definitions noted above—*user satisfaction*—is elevated, and Donoghue expects these demands for increased satisfaction and enjoyment to increase (51, 61). This is not surprising for a concept in which ease is fundamental: Donoghue's extreme usability represents the culmination of the "comfort in use" trend that began with household products contemporary to Christine Frederick.[5]

A second example shows that Donoghue's extreme usability seriously undermines the power and agency of the customers it is supposed to benefit. A clown employed by a furniture store strolls up when Donoghue and her husband are browsing and amuses their toddler long enough to allow a salesperson to negotiate a price and "close the deal." For Donoghue, the clown's role in her experience is ideal because it "was something that was implicitly understood in the way the experience was architected—not forced or artificial, but translucent, as though it was a natural part of the experience narrative" (50). However, the process is definitely *not* user-centered: "Like the clown who appears exactly when you need him, the experience will know what you need before you do" (51). The example exposes a tension in Donoghue's book: on the one hand, she advocates an approach that cultivates trust, but on the other hand, if businesses "know what you need before you do," the needs of the businesses will shape the user experience more than the user. Obviously, in such a relationship, there is room for businesses to shape "needs" to be congruent with profitability, but little room for consideration of user-centering, or the larger issues of culture I will discuss momentarily. While *Built for Use* devotes considerable attention to development of trust and desire for long-term relationships, portraying the user experience in this manner shows different ends. Later, when Donoghue argues that "[u]sability will continue to become a focus of the marketing group and will probably end as a function that lives inside the marketing group," this trust appears limited by service to profitability (203). The demand for a friction-free user experience becomes little more than facilitation of the *creation* of customer "needs," and usability is limited to buying and selling. Again, this extreme concept of usability is shaped primarily by the ideologies of consumer culture.

Like Nielsen, Donoghue emphasizes the first user experience impression, comparing it to a first date. This reflects the focus of extreme usability on novice, one-time users, and the scant attention devoted to more skilled users, or to novices who repeatedly use a certain interface. Johnson's *User-Centered Technology* points out that novice/expert separation is deeply embedded in the Western model of technology; as I note above, this separation appeared in the proto-usability developed in consumer culture, and is characteristic of the ideology of ease. Like Dvorak, who blasted the user-friendly Macintosh, some still see usability as a crutch, necessary only for less capable users (e.g., women and children). Johnson argues that novice users supposedly require ease, as designed by experts, because of their shortcomings:

> Users reside on the weak side of the idiot/genius binary. We have embedded the notion of technological idiocy so strongly in our culture that we actually begin to think of ourselves as idiots when we encounter technological breakdowns. Experts are the ones who "know," so we let them have the power, which of course means we accept whatever is given to us. (45)

This is exactly the methodology Donoghue identifies: The experience, as predefined by the expert, knows better; it knows your needs before you do. Extreme usability corrects the shortcomings of novices, enabling them to function despite their "technological idiocy." Depressingly, many people buy into this perspective, demanding extreme usability and seeing products or systems that lack it as too sophisticated to be understandable. In this manner, the frictionless and transparent nature of extreme usability becomes self-perpetuating; because novice users develop only instrumental knowledge of a system, never conceptual knowledge, their need for extreme usability—and their need for the system to know their "needs"—can be perpetual.

Donoghue concludes *Built for Use* with a look to the future that includes extensive discussion of "ubiquitous computing," the idea that computers will be spread through all parts of our culture by miniaturization and wireless networking (188–89), even embedded in human bodies, bridging the "wet-dry interface" (237). This concept of ever-increasing technological sophistication follows the uncritical concept of technology advanced by demands for ease:

1. Technology is not difficult intrinsically; most technologies can be made easy by expert designers.
2. When technology is not transparent, it has failed or is deficient.
3. Technological systems that mimic natural patterns are best.
4. Economic growth and technological progress correlate.
5. Technological progress, and the drive toward ubiquitous ease, is inevitable and natural.

In Donoghue's repeated calls for a transparent, invisible, and "frictionless" user experience, technology works best as the silent servant of humanity, and humanity works best when seeking technological growth. Extreme usability perpetuates this view, discouraging awareness of the effects of technology, unlike more complex visions of usability, which, through testing and other mechanisms, position users as critical agents who shape technology through its use. Donoghue supports understanding the needs of the customer, but radically limits this knowledge, assuming that users are novices who prefer extreme usability and that system designers envision users as critically disengaged consumers and purchasers.

As was the case with Nielsen, for Donoghue, electronic commerce and "useful tasks" are the sole end of the Web—indeed, of most technological systems. While Donoghue's approach includes examples that often entertain, the focus is buying and selling online. There is little room for other situations where considering usability might be relevant—such as the nurse-patient relationships that Mirel discusses (165–167), or the operation of everyday technologies like doors and light switches, which Norman convincingly argues should be designed with careful consideration of usability (87–91, 96–99). Indeed, extreme usability leaves little room for reading and writing technical communication.

Extreme Usability as a Methodology

Besides lobbying for an extreme concept of usability, Donoghue argues for easier pathways to implementing usability, so that it will be a "natural part" of the design and development process (203–204). This expectation is common. In the past twenty years, usability testing research has streamlined very labor-intensive protocols while maintaining or improving the quality of testing results. However, pressure to continue is provoking aggressive downsizing of usability testing, even the proposition that automated agents will replace human-administered testing. Much of the debate revolves around questions of scope: What should be considered "up for changes" during usability testing? Should evaluators restrict themselves to the most direct involvement with the product, task, or communicative expression in question, excluding environmental and cultural factors? Here the effects of extreme usability become recursive: By advancing a concept of usability shaped by the ideology of ease, the methodologies of usability come under attack as well.

Attempts to streamline usability methods are usually well-intended. Usability advocates, and technical communicators like Mirel and Johnson, have recognized the extremely problematic nature of technical systems and communication developed by system-centered methods, which all but ignore usability. Well-implemented, usability testing gives end users a voice in the design, production, and use of communication—and possibly a way to affect the practices and power exchanges that accompany its use. Fully developed concepts of usability allow immediate and more distant cultural factors to influence the design or writing process. However, though the desire to make usability testing less expensive and time-consuming is often intended to encourage wider application of usability testing and user-centered thinking in general, sometimes the result is simplifications of usability methodology that parallel the extreme usability I have already discussed.

Usability Methodology and Automatic Usability

Nielsen pioneered movement toward less difficult and costly achievement of usability. In the late 1980s and early 1990s, he published a series of papers advocating "discount usability engineering." At the time, usability testing practices mandated expensive laboratory equipment and specialized knowledge; for many, this was considered an expensive extravagance. Contemporary usability advocates took great pains to justify the high cost of their work (e.g., Dumas and Redish 18–20). Nielsen observed this could "intimidate" would-be testers, forcing them to "abandon usability altogether" (*Usability Engineering* 17). As an alternative, he suggested a less complicated methodology that reduced the number of test users involved, used more common infrastructure, and substituted less expensive procedures while still providing excellent benefits. These practices began to change the perception of usability and laid the groundwork for more widespread application of user-centered methodology.

However, Nielsen's *Designing Web Usability* took methodology a step backward. Overall, the book is a collection of best practices for web page design and information architecture based on the contemporary technological makeup of web servers and browsers. Attaining usability through iterative user-centered development is all but ignored in *Designing Web Usability*. The implied methodology—selecting from a list of techniques—is disturbingly reminiscent of outdated conceptions of design and communication, in which "front ends" are established for computer programs by adding widgets selected from a palette of off-the-shelf standards, or style is added to writing by the manipulation of surface-level features.[6]

In much the same way that a concept of extreme usability developed from application of a limited concept of usability, a methodology of extreme usability is developing by selectively applying "discount" usability testing methodology. Practitioners are looking for deeper discounts—simpler, easier, faster, and cheaper: "Discount usability techniques can be used to test a site's users without setting up a state-of-art usability lab. The methodologies are also simple and easy to implement, and the test can be completed in a short period of time, which puts discount usability well within the reach of those who can't afford the time or money to commission professional laboratory usability studies" (Kheterpal). This image of extreme usability methodology follows the conceptualization of technology common to ease noted earlier: If advancements in technology make things easier, shouldn't progress make "making it easy" easier as well? The expectation is that usability methodology, like technology, should naturally get better over time, and that complicated usability methods, like complicated technological systems, are unnecessary obfuscations or outdated relics that can be set aside. Doesn't it seem a bit odd for a methodology that includes validation of simplicity, ease, and satisfaction to be complex and theoretical?

Part of the change in methodology comes from a lack of knowledge of usability theory. Existing textbooks bolster the belief that theory is much less important than practice. Mayhew introduces *The Usability Engineering*

Lifecycle with a nod to its practicality: "This book is meant for practitioners. It is not a theoretical book, but a practical book that attempts to teach concrete, immediately usable skills to practitioners in product development organizations" (xii). Pragmatic definitions are reinforced by the hands-on nature of usability testing. Notably, focusing on practical matters is a well-established characteristic of ease. The opposition of practice and theory is often cast as "concrete or abstract," demonstrating clearly the marked term. Like ease, extreme usability methodology emphasizes pragmatic knowledge and highly specialized skills rather than generalized theoretical understanding. Results are achieved with expediency, meeting the demands for transparency and effortlessness in Donoghue's characterization of extreme usability.

Mayhew's work also reflects the desire to streamline usability testing using "quick and dirty" methods that substitute for more rigorous and expensive techniques (xiv). These "shortcuts" are intended to ensure that usability testers complete every step in the usability engineering lifecycle Mayhew advocates—even if variation in individual steps results in "varying degrees of accuracy and completeness" (20). Those using Mayhew's framework are told: "Use the more rigorous and accurate techniques described or referenced in this book when you can. But don't hesitate to use the shortcut techniques also described for any given task when necessary. They are always better than skipping a task altogether" (21). However, Mayhew provides little direction about judging the appropriateness of these "quick and dirty" techniques—perhaps as part of the desire to avoid a seemingly theoretical presentation. Though the shortcut methods are a minor part of the text (seldom more than two paragraphs per section), I am alarmed they are so integrated into the methodology. Given the pressure to cut costs, complexity, and required time, and techniques that enable such economy, I believe most readers of Mayhew's text would consider her "shortcuts" quite seriously and miss their original purpose—keeping her usability engineering framework intact. Paradoxically, Mayhew weakens her usability methodology by attempting to facilitate its complete application.

Software also shows the effect of demands for simple, expedient usability methodology. Recent versions of eHelp Corporation's RoboHelp software are marketed with continued reduction in evaluation and infrastructure costs in mind. A rich feature set makes their flagship product RoboHelp Enterprise "the fastest, easiest, most cost-effective way to create, improve and publish Help systems." It provides "comprehensive end user feedback reports that provide built-in usability testing so improvements can be made to the Help system" ("Corporate Profile"). These promises imply that performing usability testing, and interpreting and applying test results, requires minimal human involvement—instead, computers can accomplish these tasks. A white paper describing a newer version of the software more specifically names the capabilities involved:

> New feedback and reporting technologies in server-based Help can take most of the guesswork and frustration out of developing Help systems and applications. These technologies provide valuable information on Help system usage and effectiveness, and can act as a continuous usability study. Using intelligent server-based Help, technical writers and software developers have real-time access to feedback information that allows them to better understand their end users and design high-quality online documentation based on end users' needs. Software developers can use this unique usability data to determine which features and enhancements the end users really demand. ("Improving Usability" 1)

Promotional materials for RoboHelp waver on the question of human involvement: Is it needed or, in fact, even beneficial? Can it be safely omitted? eHelp marketing documents suggest—echoing the practice/theory division discussed above—that technical writers lack time to perform usability testing, and see it as a "theory" with limited effectiveness. In contrast, RoboHelp products provide "a more scientific approach to development," which is additionally more cost-effective and completes much of the work of testing on its own ("Improving Usability" 5). eHelp stops short of arguing that computers can complete replace human-administered and interpreted testing, but definitely promises reduced direct involvement.

Addwise is even more assertive about its WebArch "automatic usability analysis," claiming its software "helps companies maximize the commercial and informational benefits of the Internet by eliminating the inefficiencies and customer-discouraging flaws in their websites." WebArch gives its users "insight into actual user paths, automatically identifying flaws in information architecture" ("Addwise Services: Usability Testing"). Associated marketing materials reduce the definition of usability to effectiveness, efficiency, and satisfaction, and portray human-administered usability testing—even the "discount" alternatives suggested by Nielsen and other writers—as unfeasible and time-consuming ("Automated Usability Analysis" 10). Addwise even suggests that usability evaluators are not objective enough to be reliable—unlike their software, which generates statistically verifiable data ("Automated Usability Analysis" 8, 10).

In the aggregate, from the earliest days of usability engineering, through Nielsen's "guerrilla human computer interaction," to the partial or total automation offered by eHelp, Addwise, and similar firms, a gradual yet continual simplification of usability testing and implementation methods becomes apparent. In many cases, these changes are proposed in the interest of user advocacy—with the assumption that any attention to usability is better than no attention. But extreme development methodologies that transform usability testing into checking for best practices and adherence to conventions deemed "usable," or all but eliminate user involvement, making truly user-centered methodologies impossible, are potentially serious erosions of usability that reinforce problems caused by extreme definitions of usability.

Usability Methodology and the Question of Culture

Usability methodology is made easier not only by reducing its cost, time, and complexity, but by restricting or excluding consideration of cultural forces from usability testing and assessment. As Bernadette Longo argues in another essay in this collection, technical communicators have often narrowed the scope of inquiry through "a limited view of culture" (113; 55 in original). This approach allows us to consider the work environments or interpersonal relationships of a single organization, but prevents establishing connections to larger cultural forces that may be more forwardly politicized. A narrow perspective also discourages questioning assumptions about communication and technology that are reflected from culture into the environment, relationships, or organization under study.

However, many usability practitioners forbid even this restricted sense of culture, insisting that the technological system at hand be adapted to mitigate any problems created by the environment. In a hypothetical report based on a real-world case, Mayhew writes:

> It is not our intention to be critical of or make recommendations for change regarding either the physical environment or the corporate culture—this is beyond both our expertise and the scope of the project. We make these observations simply because any computer system will be introduced in the context of both the physical and sociocultural environment, and these factors will heavily influence its reception and usage. We cannot change the environment, but by being aware of it, we can design the user interface to any on-line system in a way that best addresses the unique requirements of its users in their environment. (110)

Why should usability evaluations be interpreted and applied in this restrictive fashion? If the "physical and sociocultural environment" can "heavily influence" communication or technical systems, why not consider changes to the environment—at very least the physical one? In the example accompanying this quotation, the environment in a police precinct was "oppressive," with harsh lighting, inadequate air conditioning, a high noise level, and other intensely negative factors (108). The evaluation admits these conditions adversely affect human performance (109), noting how this complicates the usability task at hand. Mayhew is right that proposing correctives for these conditions might exceed the expertise of a usability professional—but attempting to ignore them seems short-sighted. Why not suggest addressing some of the problems with the physical environment? Why not point out the decreased morale caused in part by the work conditions? In this case, the terrible environment sets usability up for a failure, since the proposed technical system or communication must, at least to some extent, make up for environmental defects. This is a tall order.

As I noted in my introduction, Longo points out these self-imposed restrictions occur in part because technical communication downplays the work of culture, assuming that its effects are "natural states of affairs, invisible relations enmeshed in intricate webs of institutional influences that appear inevitable" (123; 65 in original). Extreme usability's terminology and conceptualization of technological development both reflect the assumption that culture is too complex to understand. No matter how broad the concept of culture at work in a given methodology, cultural forces writ large and small are ignored because they can be difficult to change, and it's easier for all involved to go along. In a critique of Nielsen's limitations of scope, John Rhodes writes:

> I might be wrong about Nielsen's estimates but I *do* know that change is painful. Radical changes are radically painful. Many employees would prefer to suffer with what they know than benefit from change. Never forget that improving intranet usability is also about changing company culture. As many people know, **changing corporate culture is like trying to push a bus through a garden hose.** (emphasis in original)

If the intent of usability is the development of user-centered technological systems and practices of communication, then despite its difficulty, we need to engage culture, at both the level Mayhew refuses and the larger level Longo discusses. We need to consider the constructed nature of discursive practices, and "add discussions of power, politics, ethics, and cultural tensions to our understandings of what it is we do when we communicate" (Longo 127; 69 in original)—and when we attempt to improve communication and technological systems using usability.

Several specific trends in usability merit further discussion apropos the question of culture. First, usability testing often involves usability laboratories, specialized facilities designed to facilitate observation of the use of a technical system or document. In many labs, one-way mirrors conceal observers from the test user (Dumas and Redish 383–395). Usability labs can reproduce cultural disconnectedness by eliminating the context of situated use critical for usability (Mirel 182–183) and user-centered methodologies in general (Johnson 33–37). Interestingly, discount methodologies encourage portable or temporary usability laboratories that are set up in work environments as a way to reduce the cost of usability testing infrastructure (Nielsen, *Usability Engineering* 205–206). Carol Barnum also presents several methodologies for testing usability without dedicated facilities (18–21, 94–102). These methods prevent the exclusion of environmental and cultural factors removed by a quiet, discrete, isolated laboratory. Though testing outside a dedicated facility has potential drawbacks, situated testing provides the context of use called for by Mirel and Johnson and allows cultural factors to be addressed by usability evaluators.

Second, pressures to make usability quantitative and measurable have contributed to the exclusion of "subjective" factors like culture and environment. Dumas and Redish state, "A key component of usability engineering is setting specific, quantitative usability goals for the product early in the process, and then designing to meet those goals," since the development of quantitative goals is one of the best ways to ensure a user-centered development process and the mobilization of complete methodologies (11–12). But again, extreme usability is selective: as first noted above, eHelp markets RoboHelp by transforming this call for quantitative goals into a perceived need for "more scientific" methods of usability evaluation. From Christine Frederick forward, the history of ease and usability shows a continuous desire to appeal to science and engineering through quantitative measurement. Extreme usability highlights this function, shaping definitions and methodologies to eliminate attention to "subjective" cultural forces more directly associated with distribution of socioeconomic resources and power. As Longo writes, this approach is appealing because "scientific discourse is seen as producing objective truths, [but] not seen as participating in culturally contextualized contests for dominance and power" (65). This brings me to my third and final point: the exclusion of cultural factors reinforces the invisibility of power relations associated with the perspective on technology and ease at the heart of extreme usability.

Extreme usability empowers the designers or writers of a technological system or communication, excluding supposedly irrelevant pressures like culture or environment because the power of the technological system surpasses these distant forces. Johnson shows that this systemic increase in power shatters the power and agency of the end user, concentrating expertise and knowledge in administrators and designers. Rather than end-user needs, system stability is the focus, since (supposedly) end users cannot understand the complexity of technological system design (26). Dividing people into novice users and expert designers disconnects all but the most local contexts and portrays technology in isolation as "mere interaction between a user and a technological artifact" rather than including culture, history, and community: "a complicated set of social, technological, and knowledge interactions that are difficult to decipher" (Johnson 57).

Longo observes, quoting Michel de Certeau, that the scientific and practical perspective of technical communication is intended to regulate naive "know-how" (117; 59 in original). Johnson's corrective for the novice/expert binarism of the system-centered model seeks a middle ground: recovering a concept of *metis,* or cunning intelligence, which finds value in the practical "everyday know-how" repressed by science, while situating his "user-centered rhetorical complex of technology" in a field of pressures and constraints that includes explicit consideration of cultural and historical factors (38–39). For Johnson, attention to the practical functions best when balanced by the broad perspective of the cultural. Similarly, usability must acknowledge the situated character of the practical knowledge that is its focus, paying attention to culture by considering some or all of the five elements of cultural studies Longo specifies: discursivity, cultural context, historical context, implied order or subjugation, and self-reflexivity (124–127; 66–69 in original). If we fail to explicitly acknowledge culture in our definitions of usability and usability methodologies, we risk giving back recent gains in the popularity of user-centered development attributable to careful applications of usability.

EXTREME USABILITY AND TECHNICAL COMMUNICATION

While the presence of extreme usability in technical communication is now limited, the huge influence of Nielsen's *Designing Web Usability* has led some writers to advance questionable concepts of usability. In *Web Word Wizardry,* Rachel McAlpine quite literally reproduces the belief in autonomous technology encouraged by extreme usability, frequently portraying the development of good writing as a matter of magic and luck. Similarly, Martha Sammons, in *The Internet Writer's Handbook,* shows a huge debt to *Designing Web Usability,* echoing its suggestions about usable web design and "good writing" with little alteration. Both texts advance very limited definitions of usability, repeating Nielsen's best practices approach and making little mention of usability testing or other methods for measuring usability that involve readers or website users.

Thankfully, there are good models for integrating usability in technical communication. Barnum's *Usability Testing and Research* begins by historicizing usability, provides a strong theoretical argument for a multipart definition of usability, and throughout insists on a fully developed usability testing and evaluation methodology. In *Technical Communication: A Reader-Centered Approach,* Paul V. Anderson pairs usability with persuasiveness as part of a comprehensive user-centered approach, reducing the possibility that student writers will employ extreme concepts of usability that focus solely on ease and expediency. While there are problems with Anderson's text—his definition of usability is a little thin, and ethical issues sometimes have weak connections to the rest of the text—his description of usability testing methodologies concisely presents "discount" methods without dumbing them down (360–372). Technical communicators should continue to integrate usability in textbooks and other instructional materials, maintaining involvement in the process of refining and evaluating usability theories and practices. In this manner, we can reduce the impact of the problematic trends in extreme usability I describe in the last parts of this essay.

The heart of any response to extreme usability is insistence on fully developed models of usability and rejection of selective, patchwork concepts of usability mobilized to make the achievement of usability equivalent to "making it easy"—or in the service of very dubious ends, such as production of uncritical consumers in the hopes of increased profitability. Many of the definitions I mentioned earlier are ideal: Mirel argues for "a comprehensive vision of interrelated usability dimensions" that will "show that partial usability is no more favorable to users than partial system performance" (168–169). Her focus on usefulness is designed to prevent the obsession with ease of use that Quisenbery and I both see as problematic. Though it predates the explosion of the Web, Nielsen's *Usability Engineering* includes complete concepts of usability and carefully designed methodologies.

Our discipline should keep careful tabs on the pragmatism associated with usability, since the former, when associated with ease, can be incredibly corrosive. As Quisenbery demonstrates, usability has changed considerably since its introduction as a component of "human factors engineering," and a highly task-oriented approach discourages its application in situations that are not task-structured (82). In this instance, cultural studies is especially valuable, since its broad approach prevents the task orientation of usability from myopic exclusion of discursive, historical, and cultural contexts. Cultural studies demonstrates the relevance of considering the history of usability and provides strong justification for broadly construed and implemented concepts of usability. Additionally, paying attention to cultural factors strengthens usability by allowing practitioners to account for and address troublesome environmental factors as part of usability methodology.

When evaluating technical communication for usability, we should never take a "best practices" approach. This literally transactional method, which suggests that finding a path to usability can be pulled "off the shelf," is simply incompatible with user-centered methodologies. Additionally, getting locked into particular best practices shuts off usability from new forms that organize and present information in innovative ways. For example, weblogs contradict conventional guidelines for creating usable online text—yet are wildly popular and clearly very usable. A best-practices perspective could not address the usability of weblogs, since their defiance of the triad of "brief, scannable, and objective" removes them from the lexicon of extreme usability. We can more effectively understand and improve these new forms (and the online cultures developing with them) if we consider them outside of the transactional framework of extreme usability.

A cultural studies approach to technical communication shows the limitations of extreme usability as well as the strengths of the more complete concepts of usability from which extreme usability is derived. While perhaps more troublesome to teach and to implement than the streamlined concept of extreme usability, broader concepts of usability and more complete methodologies for achieving it promise much better results. The original purpose of usability—supporting and enhancing user-centered creative processes—can be achieved only by rejecting the reduction of usability to various forms of ease and making it easy. Technical communicators cannot afford to allow extreme concepts of usability to undermine the continued growth of user-centered processes of writing and design. And we should not resist the urge to point out that the ideological foundations of extreme usability have implications for our culture at large—not just for our textbooks and classrooms.

NOTES

1. Gould and Lewis offer three principles to encourage the development of usability (300). Dumas and Redish also offer a multipart definition (4). For more examples and excellent discussion of the definition of usability, see Barnum 1–12 and Johnson 80–84.

2. Examples of "extreme" being used in this way are widely diverse, including sports like snowboarding and freestyle bicycle riding, Apple AirPort Extreme networking hardware, "Extreme DVD" editions of *Terminator 2*, Aquafresh "Extreme Clean" toothpaste, and even a Jeep dealership in McHenry, Illinois. Some writers, including Donoghue, also draw parallels between the user involvement and iterative development of extreme programming (XP) and extreme usability (177, 186). See Belotti and Bankston et al. However, I believe this comparison owes more to the current popularity of XP than actual correlation of methodology.

3. Online usability forums and the weblogs of usability professionals show extensive discussion about the incongruities of Nielsen's previous work and *Designing*

Web Usability. For many people, the book diminishes Nielsen's considerable contributions to usability.

4. See chapter four of *Ease in Composition Studies* for extensive discussion of the connection of ease and writing made in nineteenth-century American composition classrooms.

5. Enjoyment and satisfaction have defined ease since the seventeenth century. See chapters 2 and 4 of *Ease in Composition Studies* for more history of ease.

6. In the preface to *Designing Web Usability*, Nielsen promises to address methodology in a second book which, as I write this, has yet to be published.

Works Cited

Addwise Corporation. "Automated Usability Analysis." 15 Dec. 2003. http://www.addwise.com/docs/WebArch-v6.ppt.

———. "Addwise Services: Usability Testing." 15 Dec. 2003. http://www.addwise.com/htmls/usability.htm.

Anderson, Paul V. *Technical Communication: A Reader-Centered Approach*. 5th ed. Boston: Thomson-Heinle, 2003.

Apple Human Interface Guidelines. New York: Addison-Wesley, 1987.

Bankston, Arlen. "Usability and User Interface Design in XP." 15 Dec. 2003. http://www.ccpace.com/resources/UsabilityinXP.pdf.

Barnum, Carol M. *Usability Testing and Research*. Allyn & Bacon Series in Technical Communication. New York: Longman, 2002.

Belotti, Victoria, Nicolas Ducheneaut, Mark Howard, and Ian Smith. "Seven Slides and a Fight: How Extreme Programming Improved Our User-Centered Design Process, but Not Our Social Skills." 15 Dec. 2003. http://murl.microsoft.com/LectureDetails.asp?966.

Dilger, Bradley. "Ease in Composition Studies." Diss. U of Florida, 2003.

Dillon, Andrew. *Designing Usable Electronic Text: Ergonomic Aspects of Human Information Usage*. London: Taylor & Francis, 1994.

Donoghue, Karen. *Built for Use: Driving Profitability Through the User Experience*. New York: McGraw-Hill, 2002.

Dumas, Joseph S., and Janice C. Redish. *A Practical Guide to Usability Testing*. Norwood: Ablex Publishing, 1993.

eHelp Corporation. "Improving Usability with Intelligent, Server-Based Help." 15 Dec. 2003. http://www.ehelp.com/downloads/serverbasedhelp.pdf.

Frederick, Christine M. *Household Engineering: Scientific Management in the Home*. Chicago: American School of Home Economics, 1919. *History of Women* 7381. Woodbridge: Research Publications, 1976.

Gelernter, David. *Machine Beauty: Elegance and the Heart of Technology*. New York: Basic Books, 1998.

Gould, John D., and Clayton Lewis. "Designing for Usability: Key Principles and What Designers Think." *Communications of the ACM* 28.3 (March 1985): 300–311.

Gray, Wayne D., and Marilyn C. Salzman. "Damaged Merchandise? A Review of Experiments That Compare Usability Evaluation Methods." *Human-Computer Interaction* 13.3 (Fall 1998): 203-261.

Intel Corporation. "Pentium 4 Datasheet." Document 298643-011. November 2003. ftp://download.intel.com/design/Pentium4/datashts/29864311.pdf.

Johnson, Robert. *User-Centered Technology: A Rhetorical Theory for Computers and Other Mundane Artifacts*. Studies in Scientific and Technical Communication. Albany: SUNY P, 1998.

Kheterpal, Suneet. "Usability On The Cheap." 3 May 2002. 15 Dec. 2003. http://www.sitepoint.com/print/741.

Longo, Bernadette. "An Approach for Applying Cultural Study Theory to Technical Writing Research." *Technical Communication Quarterly* 7.1 (Winter 1998): 53–73.

Maxwell, Lee. *Save Womens Lives: History of Washing Machines*. Oldewash Press, 2004.

Mayhew, Deborah J. *The Usability Engineering Lifecycle: A Practitioner's Handbook for User Interface Design*. San Francisco: Morgan Kaufman, 1999.

McAlpine, Rachel. *Web Word Wizardry: A Guide to Writing for the Web and Intranet*. Berkeley: Ten Speed Press, 2001.

Mirel, Barbara, and Leslie A. Olsen. "Social and Cognitive Effects of Professional Communication on Software Usability." *Technical Communication Quarterly* 7.2 (Spring 1998), 197–221.

Mirel, Barbara. "Advancing a Vision of Usability." *Reshaping Technical Communication: New Directions and Challenges for the Twenty-First Century*. Ed. Barbara Mirel and Rachel Spika. Mahwah, NJ: Lawrence Erlbaum Associates, 2002. 165–87.

Morkes, John, and Jakob Nielsen. "Concise, SCANNABLE, and Objective: How to Write for the Web." 15 Dec. 2003. http://useit.com/papers/webwriting/writing.html.

Nielsen, Jakob. *Designing Web Usability: The Practice of Simplicity*. Indianapolis: New Riders, 2000.

———. Interview with Kevin Yank. 6 Nov. 2002. 15 Dec. 2003. http://www.sitepoint.com/article/922.

———. "Report from a 1994 Web Usability Study." 1997. 15 Dec. 2003. http://useit.com/papers/1994_web_usability_report.html.

———. "The Ten Most Violated Homepage Design Guidelines." 11 November 2003, 15 Dec. 2003. http://useit.com/alertbox/20031110.html.

———. *Usability Engineering.* San Diego: Academic Press, 1993.

———. "User Empowerment and the Fun Factor." 7 July 2002. 15 Dec. 2003. http://www.useit.com/alertbox/20020707.html.

Norman, Donald. *The Psychology of Everyday Things.* New York: Basic Books, 1998.

Quisenbery, Whitney. "The Five Dimensions of Usability." *Content and Complexity: Information Design in Technical Communication.* Ed. Michael J. Albers and Beth Mazur. Mahwah, NJ: Lawrence Erlbaum Associates, 2002. 81–102.

Rhodes, John S. "Spanking Jakob Nielsen." Webword.com. 10 Nov. 2002. 15 Dec. 2003.

Sammons, Martha C. *The Internet Writer's Handbook.* Boston: Allyn & Bacon, 1999.

Watkins, Evan. *Throwaways: Work Culture and Consumer Education.* Stanford: Stanford UP, 1993.

Chapter Three

The Phantom Machine

The Invisible Ideology of Email (A Cultural Critique)

Myra G. Moses and Steven B. Katz

> • MANAGE
> • SERVE
> • PRODUCE
>
> —*E-Mail: Communicate Effectively*

> We have access to online terminals,
> but the lines don't go where we want to go.
> We are become silence, trapped by open spaces.
>
> —Steven B. Katz, "Ghosts of Technology"

Communication situations, practices, and technologies are social constructions that become invisible as they move from categories of nominal experience. Even those partially or wholly dependent on technology disappear as they are regarded as naturally occurring phenomena in a socially, culturally, and economically constructed landscape. Once naturalized, communication technology becomes as seemingly innocuous as nature (Habermas, "Technology" 86; cf. Marx). This also applies to email as a communication technology. In our society, email has become such a common and expedient medium of communication in both workplace and leisure space that the virtual boundary between work and leisure, has for all intents and purposes, collapsed. Professional life fluidly moves in and out of personal life, personal life in and out of professional; work dissolves into home, and home dissolves into work. As Judith Yaross Lee states in her study, "E-mail . . . presses against the boundaries of work and play. In a culture that has seen the work week increase and leisure decline, this paradox of electronic communication points to one of contemporary life's great ironies: Labor-saving devices make more work" (324). We will demonstrate that for today's users, email is both a professional field and a playing field where work and recreation occur simultaneously,[1] and thus in our article we widen the definition of technical communication to include personal as well as professional email.

To the current generation, email appears to be the most efficient and quickest mode of communication. Email gives us the ability to send instantaneous text messages, and to send and receive more messages on a daily basis to numerous people simultaneously, and at any time from various locations. The growth of email is obvious worldwide. After the first email message was sent on a network in 1972, only a few researchers and academics used email. In 2001, 87 million Americans were using email, and it is estimated that in 2006, 140 million Americans will be using it (Festa). In fact, for many email may be not only the fastest and most efficient mode of communication, but it also appears to represent "freedom"—both from physical location and the limitations of face-to-face communication (see Strate et al., 139; cf. Thornton 17–18 on the dominance of the United States on the World Wide Web), and from the layers of constraints and conventions that have grown in and around written discourse. In email, it seems, even business email, almost anything goes.

The freedom from physical location is evident in studies that discuss how people can email others around the world at any time (Koku; cf. Thornton), and how email allows (indeed, encourages if not compels) people to work more from various places and not be tied to the office (Kanfer; Sproull and Kiesler). As we will discuss, the unprecedented informality email allows has led both users and scholars to conceptualize email as free of traditional conventions that hamper communication via business letters. The informality of email and the breakdown of traditional conventions seem to denote that email communication is less rule-bound and more "natural" (closer to spoken language [see Lee]) than traditional communication.

The technical possibilities of email insinuate that it is a freer way to communicate (and freer communication is believed, even by social critics, to be the only possible foundation for the revitalization of democracy [Habermas, *Structural*; cf. Thornton]). Because of these technical freedoms, email has been described as a social panacea. Several studies show that some people who normally would not participate in face-to-face discussions will participate in discussions via email (Finholt and Sproull; McCormick and McCormick; Sproull and Kiesler). Email provides a seemingly "secure" space for people to contribute to discussions, enabling them to participate on a more equal level, free from social domination (Boshier; Selwyn, and Robson). Email seems to eliminate at least some social inequities; it is not always easy to determine factors such as rank, gender, power relations, and so on often evident in face-to-face communication (Dubrovsky, Kiesler, and Sethna; Huff and King). "Institutional factors enhance email's informality by establishing a sense of direct communication between equals... email exerts democratizing pressure as it bypasses the hierarchies of power and status that telephones preserve" (Lee 318). (Even Habermas, unrepentant critic of all things technologically rational, whose early work we will use to critique the ideology of email, seems in his later work to believe in the redemptive power of communication as the salvation of democracy [cf. "Technology," *Structural*].)

However, as Thornton has shown, email may be anything but a panacea. In a study of the relation between the Internet and democracy, Thornton notes the social, political, and cultural optimism surrounding the introduction of the World Wide Web as part of the rhetoric of the "technological sublime" (5), including "massive social and political change... caused by the inherent technical properties of the hardware," as well as the scale of the social revolution to be brought about by the technology (6). However, she also points out that "[t]hese recurring narratives of progress primarily function to repackage existing social structures into a new technological form, endorsing current power structures" (7). In fact, Thornton points out the illusory nature of the technological sublime/the myth of access (esp. 9), and cites a number of social, political, and economic factors that undermine any democratic tendencies of the Internet, including legislation/government intervention, fraud and censorship, surveillance, and other means of exclusion, including wealth and leisure, the disparity between Internet-rich and Internet-poor countries and peoples, literacy, gender and nationality, first-world bias, the English language, and U.S. dominance of the World Wide Web (Thornton 11–19; also see Dance).

Many studies have examined the generic conventions of email without recognizing the effect of ideology on these conventions. Lee examines email as a melding of spoken and written forms. She also mentions the democratizing tendency of email to flatten hierarchies through the adoption of a standardized memo form, for example, as well as a breakdown in formal syntax and other standardized conventions that tend to re-create or reinforce traditional hierarchies in discourse (see also Herndl; Katz 1993). In their sociocultural study that maps genres of electronic interaction in the development of an organizational community of an Italian corporation, Zucchermaglio and Talamo observe that email became increasingly informal for those within the group. Neither of these excellent analyses examines the role ideology also may play—the effect of technological values on the conventions of email, and on human behavior.

Some scholarship does raise the question of the relation between email and ideology. For example, in their "Postings on the Genre of Email," Spooner and Yancey discuss ideology in the context of the questions of whether email represents a new genre: "Many... people believe that this form of communication is new, is different, and that it enacts new relationships between authors and readers. There is, in other words, an ideology already at work here, and it entails social action" (268). At least one of the authors believes that "we will see discourse communities online arrange themselves in terms of very familiar hierarchies and conventions" (270). In a response to Spooner and Yancey's article, Deborah Holdstein states:

> I question those who would assert without hesitation that email, the Net, and the Web offer us, finally a nirvana of ultimate democracy and freedom, suggesting that even visionaries . . . beg the question of access, of the types of literacies necessary to even gain access to email, much less to the technology itself. What other inevitable hierarchies—in addition to the ones we know and understand that relate to gender, power, and so on—will be formed to order us as we "slouch towards cyberspace." . . . Yancey and Spooner's essay, then, far from and yet inclusive of its incisive concerns with genre, also helps us see the Net with a renewed, harsh glare towards the interface, highlighting to the profession the social, ideological, and power relationships replicated on new technologies and the ways in which we must acknowledge, confront, and—is it even possible?—redefine those spaces. (283-284)

While we acknowledge the obvious and unprecendented growth of email, to ignore the ideological context and content of email is to miss important cultural, political, and economic dimensions of this technological phenomenon. While admitting our own addiction, we are going to take what may be received as an unpopular position. In this article, we will not laud the social, economic, and technical benefits of email, and we will not inveigh against the technology of email or any perceived social or cultural ills it may cause. Rather, we will seek to critique the hidden ideologies that underlie email as a cultural practice. If ideology is the cultural belief in what is good, right, and best (Berlin), we look at the ideology that forms our cultural predilection for the technological and capitalistic values of efficiency, speed, and productivity, rendering them what is good, right, and best. Using Jürgen Habermas's theory of communication and work, this chapter will constitute a radical critique of the characteristics, conventions, and personal relations reified by email, to reveal the technological ideology underlying them.

We use "early Habermas" because we believe that his investigation of the invisible fusion of political, economic, and technological ideologies in communication (both business and personal), and his critique of the disappearing relation between work and social interaction is appropriate—and perhaps the only way—to understand email as a cultural phenomenon. Nancy Blyler has pointed out the connection between professional discourse and social discourse, between work and life. Using Habermas's theory, Blyler argues for the need for an oppositional ideology—a communication ideal—to make technological ideology we live in opaque ("Research"). Based on Habermas, Blyler believes "that the domain of professional communication ought to be expanded to include concerns with broad social ramifications" ("Habermas" 128).

Some impacts of email on communication patterns and behavior are due to the emphasis corporations are placing not only on writing but also on email. But email also affects personal communication. Much like Sprint cell phone commercials where problems in relationships are cured by the purchase of a cell phone, the general conception is that email facilitates and enhances personal communication and even personal relations. As Spooner and Yancey state: "We need to think of cyberspace as the commodity that it is, manufactured and marketed by today's captains of industry for the benefit of those who can afford it . . . cyberspace and its equipment are created in the real world by the same socioeconomic structures that gave us the railroad, the automobile, and the petroleum industry. It is merely our place in the hierarchy that conceals the hierarchy from us" (270–271). But even personal email must be understood as a commercial product ultimately grounded in capitalistic goals.

Habermas's theory, with its emphasis on communication and work, is useful in analyzing the ideology behind email technology and its implications for changes in technical communication—person-person as well as corporate. It will be our contention that, ideologically as well as commercially, the distinction is now so small as to be hardly discernable or significant. Ideologically, work and leisure have become virtually interchangeable. But the ideological dimensions of technology are often hidden from or ignored by the people who are too busy in both their professional and personal lives to keep up with the changing technology, never mind fully recognizing and examining the ways technology begins to influence their lives (Bolter; Spooner and Yancey). In email, as in other facets of professional and personal life, technological ideology has supplanted and replaced traditional values of business and personal communication.

At Work and at Play: Inside Institutional Frameworks

For eons humans have created technologies that serve as extensions of their own functions, enabling them to do more and making it easier to complete their work. Email technology is no different; it is a technology that extends human communication functions that includes verbal communication (speech) and nonverbal communication (gestures, facial expressions [to compensate for their loss, a whole vocabulary of emoticons has been developed] [Hafner and Lyon; Lamb and Peek; Schultz; Lee]). In corporations and other communities, the same process of extension can be understood in the development of genres, which may either be adapted from existing conventions or created anew from evolving organizational or cultural needs, writer and reader responses, and technological practices, creating patterns of communication and human behavior (e.g., see Miller, "Genre"; Orlikowski and Yates). However, tools do not serve merely as extensions; they also may cause humans to engage in new "habits of behavior," including "new ways of talking and thinking," and new modes of embodiment, and to create new institutions and technology to accommodate and/or incorporate the new tool (Miller, "Technology").

In "Technology and Science as 'Ideology,'" Habermas discusses how society's acceptance of scientific and technological advances, which supposedly improve human existence, eventually forces people to focus more on work rather than on a balance between "interaction" and work. He argues that capitalistic ideology, with an emphasis on increasing productivity, is embedded in all technology and that it is the hidden ideology that forces society to focus on work and triggers changes in existing societal systems. For Habermas, it is the disappearance of the distinction between work and leisure in both the technocultural sphere and political consciousness that demonstrates the power of the ideology of technology in our public and so-called private lives (Habermas, "Technology" 107) to erase the trace of itself.[2] In fact, even while we still retain consciousness of the ideology of technology, and so have not "succumbed" to it (cf. Habermas; Ellul; MacCormac; Miller, "Technology"), the invisibility of email technology as ideological attests to the power of the ideology to conceal itself in normativeness.

A number of scholars have noted the blurring of boundaries and the breakdown of barriers between job and play, between workplace and leisure space facilitated if not created by technology. As Lee states, "The playful qualities of these messages remind us, as Meyrowitz (1985) has taught us, that electronic media blur work and play, humor and seriousness, along with generational, social, and physical boundaries" (322). One scholar, Cheryl Geisler, has investigated the use of mobile technologies, particularly PDAs (Personal Digital Assistants). Geisler found that "though originally designed as a technology of the workplace, PDAs are regularly crossing the boundary between the workplace and personal life" (1). The ideology that blurs these spaces has resulted not only in the increased use of technology (surely one of the goals of the manufacturers of such products), but as Geisler points out, in behavioral and even cultural changes in both spheres (2). "Not only does work cross over into the homespace; communication technologies like cellular phones, *email,* and instant messaging facilitate the integration of personal life into the workplace" (emphasis ours). The result? People with wireless laptops working at the beach, and at the ballpark! (Howe).

The integration of (what is regarded as) "the personal" into the workplace heightens the awareness of the boundary between the two, but also problematizes the once distinct spheres, as Geisler (2) discusses; it also makes the distinction Habermas discusses between work and interaction even more difficult to see. As Habermas, discussing Marcuse's notion of the repression of domination, states, "subjection of individuals to the enormous apparatus of production and distribution…the deprivitization of free time, the almost indistinguishable fusion of constructive and destructive social labor" ("Technology" 83), contributes to this obscurity.

Geisler finds "the moral imperative implicit in the ideology of separate spheres has remained remarkably persistent" (2), but we believe email has broken this ideological barrier and is at home in either sphere, where it remains an invisible medium of the technology, originally designed for one purpose (and based on one ideology) but now integrated into many others. While "users may choose to use their devices for boundary maintenance or boundary crossing, for segregation or integration" (Geisler 2), we argue that no such choice seems to exist in email. Although Geisler's data "does not support the claim that the use of PDA technology will automatically entail the adoption of the time management philosophy out of which the PDA developed," it provides "some support for the belief that this form of mobile technology use is part of a general social trend to view life as a project to be worked on and managed" (Geisler 7).

Clearly, the ideology of technology is tied to capitalistic goals of production, which cannot help but change lifestyles. As Habermas demonstrates, the improvement of what is regarded as standard of living is one basis of the power and appeal of science and technology (Habermas "Technology"), despite the public's awareness of and ambivalence about the control technology exerts over their life (Blyler, "Habermas" 129–131). "Technology and science become a leading productive force" (Habermas, "Technology" 104). As Geisler also demonstrates, the development of technological products becomes the capitalistic goal of technology; economic goals become linked with technological ones, so that progress in one is progress in the other. "The development of the social system *seems* to be determined by the logic of scientific-technical progress" (Habermas, "Technology" 105). Through email use, the ideology of technology, with its capitalistic values of production, is 'imported' into personal relations. To our mind, all this extends and confirms Geisler's comment that personal life is understood as "a project."

Habermas presents a model of how technologies affect the framework of societies that accept the idea of scientific-technological progress in

"Technology and Science as 'Ideology.'" He describes two possibilities structured around the difference between work and communication. (These two possibilities seem to parallel the distinction between work and play that we have discussed; ostensibly they do. However, we will argue that interaction structured around work has become the reality of both profession and leisure life.) One possibility is that technologies are absorbed into a traditional institutional framework, a social system characterized by interaction. A "traditional society" is one in which interaction is governed by social norms; it is organized by a ruling structure where local systems of power are established but are still part of a central system of power, socioeconomic divisions determine obligations and rewards, and systems of ideals (myths, religion, laws) authorize political power. Subsystems of technology and production also exist in such societies; therefore, members of the society produce products necessary to satisfy the needs of the society through work with technology, but their behavior is still governed by traditional social norms.

Thus, the labor of the society, including its use of technology, is part of the interaction within the institutional framework and is guided by socially agreed on norms. The social norms are established using a shared language, which arises from shared social contexts of members. Based on social norms, people have common expectations about how to behave in given situations. These common behaviors are learned through acceptance of social norms and through "role internalization" or imitating others, such as one would do in an apprenticeship. If members of the society deliberately do not observe the socially agreed on norms, they are subjected to penalties that also have been socially determined. The reason for establishing social norms is to create a society with a focus on the individual and to provide a structure for communication free of domination. The society continues to function as a traditional society as long as the characteristics of an institutional framework prevail, even if purposive-rational subsystems spring up within it (Habermas, "Technology" 95).

The second possibility is that the technological subsystem that exists within a traditional institutional framework subsumes the traditional institutional framework and becomes a system of purposive-rationality. These two systems are ideologically distinct, but a purposive-rational subsystem (PRS) can exist within an institutional framework or the traditional institutional framework (TIF) can exist within a purposive-rational system. In fact, according to Habermas, in the evolutionary process of industrialization, the purposive-rational *subsystem* in a traditional society supplants political, economic, social, and cultural systems of the traditional society, as illustrated in fig. 3.1. So, for example, in a traditional medieval society, the institutional structure is ruled by the divine right of kings, religion constitutes the value system, and the purposive-rational subsystem is confined to the treasury; but

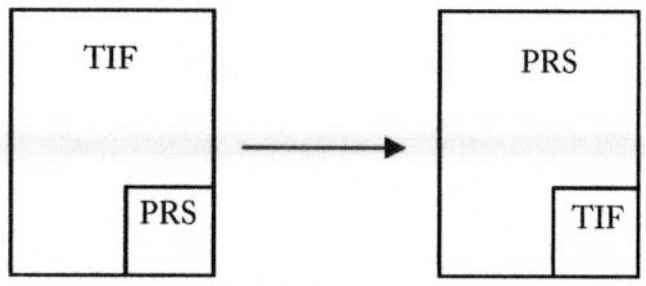

FIGURE 3.1. Schematic representation of the evolution of institutional frameworks

with the rise of the bourgeoisie, the king is beheaded, religion is boxed in a cultural corner, and the economic values of the treasury are now the dominant framework of the society.

Habermas distinguishes between the two systems in seven categories, which can be explained by our questions in the first column of table 3.1 that interpret the categories Habermas applies to both systems.

TABLE 3.1

Habermas's Chart of Work and Interaction. Reprinted from *Toward a Rational Society* by Jürgan Habermas. Copyright © 1970 by Beacon Press. Reprinted by Permission of Beacon Press, Boston © Suhrkamp Verlog, Frankfort am Main. All Rights Reserved.

Our Interpretive Questions		*Institutional framework: symbolic interaction*	*Systems of purposive-rational (instrumental and strategic) action*
What governs actions?	action-orienting rules	social norms	technical rules
What type of language is used?	level of definition	intersubjectively shared ordinary language	context-free language
What types of behavioral expectations exist?	type of definition	reciprocal expectations about behavior	conditional predictions conditional imperatives
How are behaviors acquired?	mechanisms of acquisition	role internalization	learning of skills and qualifications
What is the purpose of actions?	function of action type	maintenance of institutions (conformity to norms on the basis of reciprocal enforcement)	problem-solving (goal attainment, defined in means-end relations)
What happens when rules are violated?	sanctions against violation of rules	punishment on the basis of conventional sanctions: failure against authority	inefficacy: failure in reality
What is the reason for working?	"rationalization"	emancipation, individuation; extension of communication free of domination	growth of productive forces: extension of power of technical control

While in traditional societies the systems are governed by social norms, in technological societies the systems are governed by technical rules. The technical rules are determined by the technology itself and are based on "empirically true or analytically correct" information, and are therefore not defined by the context of traditional societies, but rather by the purposive-rational context of industrial societies in which science and technology have become primary value systems. To function in the purposive-rational system, people must follow the technical rules by learning a set of skills and qualifications. As a result, people's behavior is regulated by "conditional predications" in which the desired outcome determines their actions and also by "conditional imperatives" in which an existing condition determines what the action should be. Both conditions are created by the technology itself–or the socioeconomical and political systems required to maintain it (Habermas, "Technology").

The resulting types of conditional behavior demonstrate the means-end nature of the technology established by the technical rules. If the rules are violated, either intentionally or because a person does not have the requisite skills, the consequences, like the rules, are also determined by the technology. Therefore, if the rules are violated, the technology does not work properly and the consequence is that the person does not successfully achieve the desired "end." Since the only outcome of following the technical rules is a successful use of one's skills to achieve a goal, the underlying focus of purposive-rational systems is to increase productivity, which results in an extension of technical control. The increase in technical control and productivity does not threaten the institutional framework unless it challenges the underlying ideas and social norms of the institutional framework. Purposive-rational subsystems that function within institutional frameworks as subsystems generally do not challenge traditional systems, but instead incorporate capitalistic ideology within traditional societies to expand technical control and increase productivity. However, when technical subsystems take over institutional frameworks, societies are governed by technical rules for increasing productivity.

IN THE BEGINNING: THE MILITARY GENESIS OF EMAIL

The ideology underlying technologies can often be determined by looking at the original motivations and reasons the technologies were created, the circumstances surrounding their creation, or both. Email was not an idea conceived by a group of inventors or business investors determined to develop a mass communication technology. Rather, email developed as a result of the Defense Department's ARPANET network technology, developed by engineers at the request and expense of the United States Advanced Research Projects Agency (ARPA). The purpose of ARPANET was to allow researchers to share technical resources (computer coding/software and technical data) in order to economize and increase efficiency of the research they were conducting and speed up the production of technologies they were creating. The purpose of ARPANET did not include plans for researchers to communicate with each other using text messages (see Geise on the clash between the Department of Defense mission to improve military communication, and the desire of the computer scientists working on the technology for a freer flow of communication). People who used ARPANET could exchange messages with each other using "intracomputer mail," a technology that had existed since the 1960s, which allowed people using the same computer terminal to leave messages for each other.

"Intracomputer mail" was the only type of electronic mail available until 1971 when Ray Tomlinson developed "network mail." Tomlinson was working on a program to send technical information between computers when he decided to alter the program's code in an attempt to send a text message. His modification worked, and he sent the first email message (the content was probably "qwertyuiop") to himself from one computer to another computer, both of which were sitting side-by-side in the same room (Hardy; Hafner and Lyon). Thus, the first trial of network email was not an effort to communicate with another human, but to see if the technical rules would allow the possibility. This history of email's origin shows that although researchers did not intend to create an email system based on "rational ideology," email nevertheless grew out of the Defense Department's desire to increase efficiency, speed, and productivity of its "workers."

Since its inception within the context of ARPANET, the development of email has continued to focus on increasing speed, productivity, and efficiency, and has become driven by economic growth and the continued commercial nature of most Internet sites (Geise 150–158).[3] This is evident in the way the two leading messaging vendors, Microsoft and IBM, design their email technologies. Microsoft and IBM design their email products to enable both corporate and personal users to increase their productivity and efficiency, often in conjunction with increasing the speed of transmission and other technical factors. As Geisler suggests with mobile technologies, the technical goals and features of email can be discerned in part from the explanations and descriptions of electronic products in advertisements "for insights into how users may view devices, what motivates acquisition, and what expectations for use adopters may bring to the experience of incorporating new technologies into everyday life" (1).

The Corporate Takeover by the Email Industry

The focus of Microsoft Outlook, Microsoft's email software, is evident in the product guide: "Outlook version 2002 can help users manage their time and information more effectively" (MPG 2). The guide states the new version's goals were: "to make working with email, tasks, contacts, and appointments more intuitive without requiring users to learn new ways of accomplishing their tasks or spend time searching for these tools" (MPG 1); "to enable users to spend time working rather than worrying about their software" (MPG 9); and "to make sure setup and configuration was made simpler so that users could stay focused on being productive instead of worrying about their software" (MPG 11). In these three points, we see that underlying the effort to improve communication technology at work is a fundamental belief in the values of efficiency, speed, productivity—what Habermas identifies as the ideology of science and technology in its fusion with capitalism.

The purpose of the email technology is also evident in the list of descriptions, included in the product guide, for thirty-four new or improved features. Twenty-one of the descriptions explicitly describe how the features allow workers to complete necessary tasks such as finding messages (MPG 4) or looking up contacts (MPG 7) more "easily," "efficiently," and "quickly." The remaining thirteen descriptions are still focused on how the features increase productivity, but they are more subtle. For example, some descriptions assert that the feature would allow the user to complete tasks such as "AutoComplete Addressing" for sending email without looking for an email address (MPG 2) or finding "messages, appointments, or tasks" (MPG 4) more quickly and easily. The fact that it is technologically possible for users to complete tasks easier and faster not only culturally signifies but pragmatically means that workers can now be more productive. In fact, two of the descriptions directly emphasize that the feature keeps the users working/producing: the "Cancel Request to Server" feature—"this allows Outlook version 2002 to be more resilient to network or server disruptions and enables users to stay working" (MPG 8), and the "Document Recovery" feature—"as a result, users spend less time recreating their email messages and spend more time working" (MPG 9).

We see the same ideological impulse at work in the advertisement for IBM's email software, Lotus Notes. In an article describing the impact of Lotus Notes and Domino 6 (Domino is the email software for the server), IBM states they developed the latest version of Lotus Notes based on research that revealed customers wanted browser-based email, which would allow them to check email from any location with a web browser without requiring special software. The ideology of Lotus Notes is implicit in the way IBM describes the product: "We're able to raise the level of collaboration and productivity to new heights," and "we are seeing major performance improvements and reductions in costs from 25 to 40 percent" (IBM). These descriptions also expose a view of workers as (human) resources or as capital to be purposively used or invested by their organization, or as parts of rational, capitalistic production systems ideally existing within organizations (cf. Miller, "Rhetoric"): "we're offering our customers the ability to significantly drive cost out of their organizations, to increase productivity, and to maximize their returns on investment in their people," (IBM) and "now they can get significant improvements in performance on the server side—plus a set of client capabilities that really unlock the productivity of their organization" (IBM).

These viewpoints, along with Lotus Notes' goals and product descriptions, reveal the desire behind IBM's email technology to increase productivity of businesses and employees. By designing their email software to help companies increase productivity and to help workers become more efficient, IBM and Microsoft develop products that continue to be based on the ideology inherent in email since it was created as an extension of ARPANET. According to Habermas's theory, since they have been based on increasing productivity, efficiency, and speed, the origin and the ongoing development of email technology demonstrates that the "rationality" of email is oriented toward the production of work, a means-end relation to which all other goals, including communication, are subordinated. As with ARPANET, the focus is on productivity; improved general communication is really a spin-off.

Personal Email and Technological Relations

The ideological genesis and continuing focus on increasing email's efficiency and speed is not confined only to email systems that are developed for business environments, but also underlies the commercial development of email for personal use that is easier and faster, and reshapes personal relations in ways that seem to have been unpredicted or unseen. While some people may use Hotmail (the largest online provider of free email services) for business reasons, it is currently intended more for personal communication, as evident in a Hotmail update description: "The new, more intuitive and consumer-friendly interface makes it easier for people to manage email and keep in touch with friends and family they care about most" (Microsoft 2003b). Although it is used primarily for personal communication, the same underlying ideology that exists in business email applications can be seen by examining the original goals of Hotmail creators that were the basis of the application when it was launched on 4 July 1996, and the purposes behind the extensive updates to the application that occurred in July 1999 and December 2003.

On July 4, 1996, Hotmail's creators, Sabeer Bhatia and Jack Smith, activated it as a free, web-based email system, which they had designed to be "fast, easy to use, reliable, and accessible from any Internet-connected terminal" (Microsoft 1999a). This was a deliberate attempt to create a technology that provides users with a sense of apparent freedom in that Hotmail is clearly intended to be available to everyone on the Internet and to be available from any Internet connection. Hotmail was to be the great equalizer of the World Wide Web.[4] The original goals remained a fundamental element behind Hotmail's design even when in July 1999, three years after launching the service, Don Bradford, Hotmail's general manager, announced the first major update to the email application, "to provide service that is fast and reliable, with easy-to-use features that help our members get things done online" (Microsoft 1999b)—that is increase speed, productivity, and efficiency, repectively.

Thus, although Hotmail is not intended primarily for business purposes, these press releases hint at an ideology of increasing productivity—even if it is the productivity of leisure. What is no longer obvious, however, is that productivity, efficiency, and speed in personal relations are technological values—ideological goals imported from the realm of work that turn personal relations into means-ends relations—technological relations whose very end is to make personal relations productive, efficient, and fast. Email is not primarily or only about 'reaching out and touching someone.' That is not why companies produce email products, whether corporate or personal. It is about providing a service to make a profit (Spooner and Yancey 270–271). Perhaps more important, the effect of this capitalistic ideology is to fundamentally alter, through the values embedded in the communication medium of email, the relation of *users* to machines, to each other, and to themselves, turning the purpose of that relationship into the work of technological capitalism.

An emphasis on increasing efficiency and speed as the basis for more "productive" relationships can be seen not only in the repetition of phrases and concepts such as "speed," access, and ease, but also in the new features of the updated Hotmail service, such as "faster page loading times," "easier access to tools through a new navigation scheme," more "on-screen work space" and "decreased loading times for interface" (Microsoft 1999a). These features become not only the technical means to human relations, but as we will show in the subsequent sections, ideological ends themselves: the speed, access, space, and interface establish not only the technological parameters of email as leisurespace, but also the new conventions of personal communication, and the means-ends relations created by and through them.

Although the original design and early development of Hotmail certainly incorporated the values of efficiency, productivity, and speed, the emphasis on those values is possibly even more prominent now due to Hotmail's current affiliation with Microsoft. Four features in the update that embody these values in software allow users to handle more information more efficiently, store larger amount of messages, and archive messages more efficiently. Directions for these new features shown in a flash tutorial on Hotmail's website about the update include how to: respond to MSN Instant Messenger messages from within Hotmail using "Instant Reply," enhancing the speed at which people can communicate via email applications by allowing the capability of responding to email messages with synchronous communication from within the email application itself; combat junk email by improved reporting and filters, demonstrating that Hotmail is trying to help people be more efficient by providing better ways of handling large numbers of personal email; and coordinate calendars, including updating schedules, sending and tracking meeting requests, sharing calendars, and receiving reminders (Microsoft 2003a).

These features show that personal email applications and business email applications are becoming similar since business email applications, like Microsoft Outlook and Lotus Notes, have calendar functions, and previously Hotmail did not. They also show how Hotmail imports the values and goals of business into personal life, turning life, as Geisler states in connection to similar PDA functions, into a project. For example, in the tutorial, phrases such as, "several tools to automatically save information," and "it's easier to keep track of important names, addresses, etc. than ever before—and to do more with that information," point to the desire and goal of users to organize their personal relations based on rational systems—and to do so more efficiently and quickly to increase the "productivity" of those personal relations (Microsoft 2003a). The outcome of this desire conflates expectations between business email and personal email and turns personal relations into means-ends ones. What's at stake here is changing how people relate to each other.

The Hotmail website also includes an "All about Hotmail" FAQ section that comes with a description of additional features that demonstrate further blurring between work and leisurespace. For example, users can access Hotmail from Outlook; this allows users to set up an Inbox in Outlook for their Hotmail account—a reversal of the relation of leisure and work, for here, instead of work intruding into leisurespace, Hotmail as a technology designed and used primarily for leisure communication is encouraging the encroachment of personal life into the workplace.

In addition to selling the product, the intent behind the new design is to provide users with control in order to communicate more effectively; but another part of the intent is the *management* of personal life. This is clearly stated by Blake Irving, corporate vice president on MSN Communications and Merchant Platforms at Microsoft, in regard to one of the features, "this is just one example of how MSN is providing people with the tools and technologies they need to better manage their email and, ultimately, their time"; Hotmail thus holds out the illusion of control: "The new MSN

Hotmail provides consumers with a wealth of tools to gain control over their email, calendar and contacts" (Microsoft 2003b).

We want to note that the changes in service implemented by Hotmail, while still adhering to the original goals of the creators, were partially in response to the desires of users for more "speed, ease of use and reliability" (Microsoft 1999a). Email applications used in the workplace are still influenced by these goals, goals driven by a combination of the desires of consumers, developers, and corporations. The desire on the part of consumers for more speed points to an important facet of the ideology of science and technology that Habermas discusses: the desire not only for technological goods and services but also technical relations and ends, originates, or appears to originate, at the grassroots level, from the bottom up ("Technology" 105–122). In fact, for Habermas it is the grassroots nature of the ideology, and the higher standard of living the economic surplus affords, that gives the ideology its stability and power. But as we know, the relation between ideology and the desires of users is complex, also involving economic needs, cultural goals, created expectations, and appealing products.

R&D VS. I&D

While people usually think about research and development (R&D), they don't think about ideology and development (I&D), about products as ideologies or ideologies as products. But as Bakhtin and Foucault in different ways have shown, ideologies are material, and material things have ideological dimensions and uses. Thus, perhaps we can think about ideological development in ways similar to product development. Ideological product development, which drives the ideological as well as technological development of email applications, perhaps can best be represented by fig. 3.2.

Feature Applications and Ideological Products

Email developers/providers use the term "application" to refer to email packages, software, programs, and so on. As we have depicted in fig. 3.2, *feature application* involves the design, development, testing, and release not only of tangible goods, but also of *ideological products and technological innovations.* Email applications certainly go through this process. What is distinctive about email, perhaps, is the less obvious Marxist dimension, which may be true of all services where the primary means of production remain in the hands of the provider. Email is an ideological product.

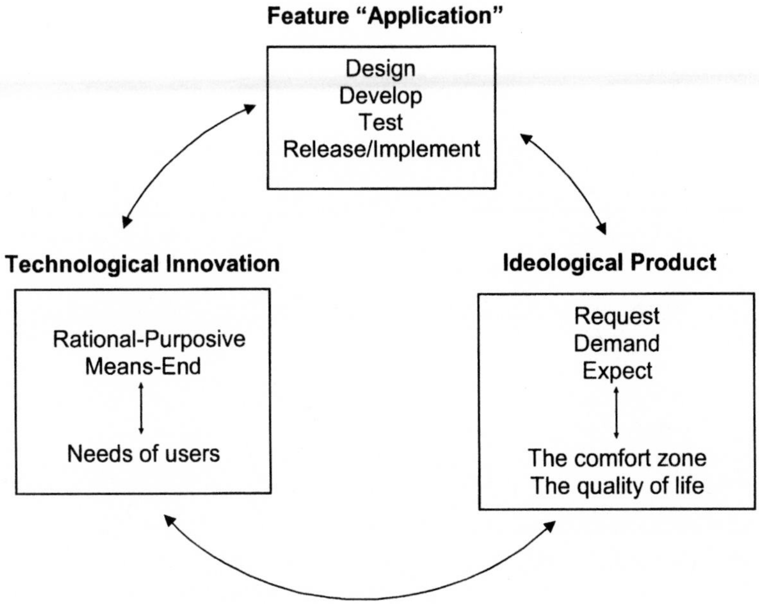

FIGURE 3.2. Ideological product development system

Although users do not really have the type of control that allows them to determine production, they do have the ability to exert some influence over the way the application is designed by letting the developers know about their expectations. In the update to Hotmail discussed above, Microsoft deliberately restructured Hotmail's interface in response to user request/demand/expectation to make it seem more like Outlook, Microsoft's business email application. Although at times the providers appear to hand control over to the users by advertising a number of new features that users asked for, this is only an illusion of control. We do not own the company, we do not own the means of production, we do not have the power to make changes; we only have the power to imbibe expectations, and make suggestions based on them. The actual and virtual service that companies provide is the illusion that we are in control of our communication and of organizing our lives.

The Ideological Product and Technological Innovation

Feature application and ideological product create and condition each other. But ideological products are in part created and conditioned by purposive-rational/means-ends relations, which therefore shape the needs of users by defining comfort zones and quality of life. Technological innovation is the interaction between means-end relations and the needs of users. The illusion of control creates expectations, demands, and requests for more illusion of control, on which the notions of comfort and standards of living are based. Within the Ideological Product box in fig. 3.2, we see the interaction of requests/demands/expectations in relation to zones of comfort and qualities of life; but as part of a larger whole we see that both requests/demands/expectations, and comforts and qualities, are the ideological products not only of feature applications manufactured by developers/providers, but that they are also the needs of users—"manufactured" by technological innovation based on the purposive-rational system.

Technological Innovation and Feature Application

The testing and development of applications creates the request/demand/expectation of users, which are in fact not only technological products but ideological products based on needs manufactured by technological innovation. But feature application is created not only by manufacturers/developers/providers, but also the purposive-rational system of technological innovation. The needs of the users spur technological innovation. But feature applications are shaped by the purposive-rational system (in a process that is predicated on means-end relations). Thus the economic and political basis of technological innovation is concealed from the users who keenly feel their need nevertheless.

An example of the system of ideological product development would be the addition of a calendar function to Hotmail because it was already included in Outlook and Lotus Notes. The ideological product is both the desire and the outcome, the means and the ends.

The In-Out Box of Institutional Frameworks

Communication, whether at work or at play, is still framed by and exists within the ideology of technology. What is significant here is the ideology of technology, not just the technology itself, or what consumers use it for. Features of email can be understood within that context. People's expectations of leisure, at least in the United States (cf. Thorton; Dance), are shaped to a certain extent by their experience of email technology, both professional and personal. Email is thus both a workplace and leisurespace technology. People come to expect certain characteristics they see as valuable in the technology used at work, which is based on principles of increasing efficiency, productivity, and speed, and therefore start to "demand" the same fast, efficient, and productive characteristics of leisure technology. Further blurring expectations, people expect to be able to perform the same tasks at work and home.

A primary purpose of fig. 3.2 is to point to how needs of users are by-products of the ideological product created by both feature applications and technological innovation. Marxists usually think about needs in economic terms of survival (cost of living), and Maslovians in psychological terms of emotional development (hierarchy of needs); but here, both economic and emotional needs are shown to be created and conditioned by purposive-rationality. In other words, needs are constrained by the means-ends relation imposed on them by technological capitalism. Within this system, accessibility of email applications is a technological reduction of the notion of personal relations (see Buber). Someone experiences a need to communicate, to which business responds: "You've got mail!"

Characteristics: Personal and Technological

Due to emphasis on productivity embedded in the goals and development of email technology, many individual characteristics of email allow for and indeed almost demand increased productivity. Such email characteristics not only have impacted the way people communicate, but they also have impacted expectations of how people should communicate with others. These new expectations are determined by technical conditions established by email technology rather than by reciprocal expectations about behavior that are based on traditional social norms. The ability of email technology to deliver messages almost instantly has changed the way people communicate in many obvious ways. People can now send messages anywhere in the world that has Internet access. As a result, the speed of email has led to behavioral expectations that are not defined by reciprocal expectations, which are established by social norms of an institutional framework. Instead, speed, which is made possible by the technical rules that are inherent in email software and hardware, has established behavioral expectations. Some of the behavioral expectations regarding the speed of email are based on what Habermas terms "conditional predictions," which is when people found behavioral expectations on the logic of: if I/they take action X, then condition Y will be the result. For example, since email is generally delivered almost instantaneously, many people hold the expectation of: if I send an email about situation A to person B, then person B will know about situation A instantaneously (an expectation that is obvious in the almost inevitable response when person B states that he or she does not know about situation A and the sender says, "but I sent you an email about it") (Dobrian; Graham).

Another behavioral expectation resulting from the speed of email is that people can handle more pieces of communication on a daily basis. Studies show the typical worker spends at least two hours daily dealing with email messages (Ferris 2000b). The time needed to deal with email will only increase as the amount of email received by corporate users increases yearly by 35 to 50 percent (Ferris 2001).This type of behavioral expectation is what Habermas calls the "conditional imperative," which is the logic of: if condition X exists, then I/they will do action Y. For example, many people expect that since email can deliver numerous messages quickly, they should be able to respond to more messages faster. Therefore, since both condition X and action Y are made possible by the technology, people come to expect it.

Other behavioral expectations that result from the technical characteristics of email can be framed as conditional predictions or conditional imperatives, such that since email is "always" accessible, people feel like they have to check email continuously, often setting up email notification systems that alert them when they have new messages (Schultz). Computer users are constantly interrupted by new email messages. The characteristic of accessibility also sets up expectations that people are always within reach, even if they are not physically at work, and should respond to email from home in the evening, on weekends, and even on vacations. These behavioral expectations are encountered frequently because they have developed and spread along with the rapid development and expansion of email use, and they shape emerging behavior patterns.

In analyzing their behavioral responses to these expectations, some people may appear to develop an "addiction" to email, also known as "emailoholism" (Tschabitscher). One symptom may be that people expect immediate responses to their messages (this is not always a reciprocal expectation); because of the instantaneous nature of email, they feel the need to send and receive messages instantly, and thus constantly check their email just in case messages are waiting, developing what Schultz calls "communication enslavement." We might define email addiction as a three-prong need based on technological values of productivity, speed, and efficiency that indemnify technological-capitalistic systems. Thus, we understand addiction as ideology. By making people feel the need to communicate constantly (productivity), instantly (speed), and directly (efficiency), email users develop communication dependence. One "addict" confesses, "I'm finally ready to admit I have a problem. At a trade show recently, I awoke in the middle of the night, unable to get my laptop 'dial-up' connected to the hotel [phone system]. I found myself in a 24-hour hotel business center looking for a 'fix'" (Weil, "Finding"). Another "addict" describes the enormous impact of email: "I have been strung out on checking my email every minute for 7 years. The issue for me comes in when I get home and have to check personal email. . . . I feel that my life is fading away" (Weil, "Finding").

Research about people's technical relationship with email supports these personal accounts of addiction. The Center for Online Addiction acknowledges that email is one of the most addictive online applications (www.netaddiction.com/whatis.htm). The Gartner Study Group provides statistics about the extent of email addiction: at the time of the report, 53 percent of business users check email six or more times on workdays; 34 percent of users check email constantly during the day; and 42 percent of American email users check email on vacation (CBSNews). These statistics are significant, but also significant are the facts that both users and researchers recognize and admit such an addiction exists, that researchers are monitoring it, and that some people even take the addiction seriously enough to suggest ways to handle/control/break the addiction (Weil, "Finding").[5]

The Ties that Bind: Conventions and Technical Rules

Email has developed quickly and continuously, and thus guidelines and conventions are emerging that are comparable to those governing business letters and memos (Benjamin; Lamp and Peek; Terminello and Reed). Due to the rapid expansion of email, people have not had time to develop rigid, social conventions for using it,[6] but the technology of email itself seems to be creating a set of emerging conventions. These new conventions seem to reflect a means-end ideology based on technical rules. An example of how these conventions are governed by the technology can be seen in formatting conventions, such as the way technical rules determine the memo header. The header section of an email message generally includes the following traditional fields: "From," "To," "Subject," "Cc." However, as Yates and Orlikowski state, "In this case, computers rather than people routed the messages, so the fields of the memo heading were designed to be readable by computers (as well as humans). . . . [S]ystem identifiers . . . are sometimes clearly recognizable variants on the individuals' names, sometimes they are nonmeaningful sequences of letters and numbers" (316, 317).

In email there is no possibility of deviation from the memo header because technical rules establish what must occur in the header. The software enters the sender's email address automatically in the "From" field, and the sender must enter an exact address in the "To" field. If these fields are used correctly, the message will usually reach the receiver whether or not the sender uses the remaining three fields: "Subject," "Cc," and "Bcc." If the "Subject" field is left blank, most email software will query the sender to enter an appropriate subject line; in a sense, the software is asking whether the sender really wants to circumvent the convention. Therefore, the sender's message is impacted by technical rules in two ways: (1) violating the technical rule does not change the "memo format" of the email because the space for the subject line is still there; and (2) violating the technical rule could result in an "efficiency failure" because the receiver has to open the message to determine the subject.

Thus, even though the email header, with the exception of the "Bcc" function, is grounded in the traditional memo heading format, the email header is not based on traditional social norms and expectations (Hafner; Hardy). The fields of the email header that the receiver sees resemble a memo (Hardy 31). Traditional norms and expectations, as evident in business communication handbooks, require the memo format to be used for internal messages and the business letter to be used for external messages (Baugh; Inkster). When the members of ARPANET first began to use email, 60–70 percent of the messages were sent internally to people. Now that email is used both internally and externally, the format of the email header is based on technical rules that have blurred the distinction between internal and external messages (perhaps explaining students' confusion regarding when to use memos vs. letters).

The result is that while email users readily adapt the memo form of email as a genre to corporate or personal ends (Zucchermaglio and Talamo), the email technology itself, not the social context or the user, chooses how the message will look. Email has apparently flattened the traditional internal and external boundaries of business communication by instantiating memo header conventions (cf. Lee 310). Other factors influence the relation of senders and receivers, but the technological embodiment of the memo heading as the generic form of email is indicative of a purposive-rational system rather than the traditional institutional system.

Technical rules have also generated guidelines regarding how email text should be formatted. While users can often determine text formatting on their machine, they are ultimately not in control of how it will look on the receiver's machine, depending on software compatibility. This compatibility both instantiates and creates the set of expectations that constitute technical rules and to some extent govern behavior. For example, some email software will automatically format the spacing of the text by allowing for text wrap and will allow the user to format text in ways similar to that of word processing software, such as using bold, italics, underlining, bulleted lists, different font sizes, colors, and types. However, the email software used by the receiver may not be able to decipher the formatted text, and the text may appear as plain, unformatted text or may include gibberish characters in place of the formatted text. Therefore, email manuals tend to stress that formatting is not important since the sender cannot control how the receiver will view the text (Benjamin; Bly).

In Habermasian terms, technical rules can be either nonoptional or optional. In regard to the formatting of email, if nonoptional technical rules in a purposive-rational system are violated, rather than the consequence of traditional, socially agreed on punishments, the "failure in reality" is a failure of message not being sent (it never reaches the inbox, or reaches the wrong box). This failure in reality could have two possible causes: a human violation or a computer violation of technical rules. A computer violation would be when a computer violates its own technical rules, for example, when it inserts "nonsense characters"—html code—into an email message. Someone receives this message: "Love?   are you kidding??" In the context of a troubled relationship, he interprets the   as "and no bullshit please," which makes all too perfect sense in the context of the email, further damaging the relationship. In actuality,   is html code for "non-breaking space," but was inserted in the message (many more times than shown here) in linguistically significant and emotionally crucial places as above by the email application. In cases like this, the phantom machine becomes visible—its code transmigrating software from the realm of the unseen to the screen.

As we see in table 3.1, a violation of "optional" rules also could result in a failure in reality. One type of optional rule is technical, as when, for example, a receiver has not turned on text wrap and is forced to use the horizontal scroll bar to read long lines of text instead of being able to see the message in one glance. The other type of optional rule is "nontechnical." As noted in many netiquette books (see Terminello and Reed 56), a sender should use a specific subject line; if this rule is violated, the receiver may not read a message or may misunderstand the content and/or purpose of a message even after reading it (cf. Lee [316] and Zucchermaglio and Talamo, who believe that there is a correlation between length and detail of subject lines and personal relations among groups of email users). While the subject line is a holdover from the traditional memo form, its incorporation in email in effect becomes a technical rule (a "soft necessity") not because it is embodied in the genre, but because its nonuse violates the ideological means-end relations in which the email message exists.

Email Styles and Personal Relations

Conventions emerging from the way users are adjusting to email technology suggest that users may be moving away from rules established within the traditional institutional framework toward a purposive-rational system.

For example, traditional rules governing the salutations of business letters are based on social norms and mirror a firmly entrenched social hierarchy, a characteristic noted by Habermas in traditional institutional frameworks. Malcolm Richardson and Sarah Liggett, discussing social hierarchies in medieval and modern letters, showed that hierarchical differences between people were especially evident in the formal wording of the salutations, which were governed by formulas dictating how a writer should address the reader based on the reader's societal position (see Anonymous of Bologna, who presented a hierarchical list of salutation formulas based on social position).

Both the advice of the email "experts" and the actual behavior of the users depict a move away from traditional forms of salutation. Email guidelines suggest including a traditional salutation in an email the first time a writer sends a message to someone—afterward the sender may use the receiver's first name or omit the salutation entirely (Angell and Heslop; Locker and Kaczmarek). According to Laurie A. Pratt, the recommendations of the communication experts are reflected in the writing practices of email users: only 3.9 percent of messages include the standard greeting format "Dear Mr./Ms. X," of traditional business letters. Instead, people tend to use a more informal salutation style (research shows 46.8 percent of messages use a more informal salutation by only including the first name of the receiver, and 48.3 percent of messages do not include a standard salutation or either the receiver's first or last name [Pratt]).

The move away from traditional conventions may indicate that new email salutation conventions reflect Habermas's idea of language free of traditional "social" context, as shown in table 3.1, since senders can usually address both superiors and subordinates by their first names in email messages without fear of reprisal based on violations of traditional social norms, and because social interactions in this case are now at least partially governed by "technical rules." The resulting communication style seems to indicate a flattening of traditional social hierarchies with less emphasis on a traditional approach where an "intersubjectively shared language reflects social norms and hierarchies" (i.e., writers addressing readers based on their title, hierarchical position, or both).[7]

In the context free category (table 3.1), another move away from the traditional institutional framework appears in email style, which often combines conventions of writing with those of speaking (Lee) and thus leads to a more informal style of writing (Zucchermaglio and Talamo 279–280). Email users pay less attention to conventions of traditional, formal business communication, such as structure, spelling, and grammar (Benjamin; Yates and Orlikowski). As Lee remarks, "[B]etween people who already know each other or who share institutional links—the vocabulary and grammar of email lack the formality of many other genres of professional communication"

(318). Other "nontraditional" email conventions include short replies that depend totally on known context, "less contextualizing or background," less redundancy; an abundance of abbreviations, use of lowercase, initials in name, phonetic spellings, jargon, colloquial vocabulary; incomplete or ungrammatical sentences and "haphazard" punctuation (309–319). An email that one of us received from a student reads: "Hey, sorry about... ill just copy paste lol...cause files are too difficult for my computer illiterate brain ☺" (mrrandal). This email message does sound more like a verbal conversation than a written message.

For Lee "deviation from formal grammar usually signifies efforts to visualize talk.... [E]-mail constitutes a junction in which orality and literacy, in their extreme or purest forms, meet. One reason is obvious: email adapts the technology of the keyboard, a by-product of print, to the requirements of talk" (319, 323).[8] One example of written and spoken conventions being combined is that in addition to incorporating and adapting written conventions regarding salutations, email also incorporates greetings used in telephone and face-to-face conversations. Some people have integrated the conversational convention of "Hello,"[9] as well as other characteristics of informal writing, into email as an informal replacement for both complex medieval and streamlined modern salutations based on rank, social class: "hello! hope this finds you sane and well....im doing great...(smile). still shying away from school...(im laughing)" (julichny). Other changes might include: rapid reply; reply with prior message included; subject lines that can be the whole message, completely dropping salutations and closings used in letters and memos (Lee 311); "[r]eceipt tracing," when the code appears as a form of closing, "confirms the end of the communication in much the same way as the blank space at the end of a letter or memo" (Lee 317).

These scholars attribute the shift to informal styles on email to the merging of oral and literate traditions, the development of personal relations on and offline, and changes in genre. Yates and Orlikowski attribute the toleration of spelling and grammatical errors in email to "the typical rapidity of and lack of secretarial mediation in this medium, as well as its weaker editing facilities and lack of typing skills among many electronic mail users" (317). But the shift in conventions and styles may go beyond the quotidian. For example, speaking of the subject line, Orlikowski and Yates state, "[I]n this case we see a creative use of a formal structure that emphasizes efficiency of the communication permitted by the electronic-mail system" (266). In fact, it is possible that the shift in writing conventions and styles is at least partly driven by technical rules and values created by email technology itself. Ironically, despite the informality of email and the apparent freedom of it, minimal context and the proliferation of abbreviations, phonetic spellings, colloquial vocabulary, and incomplete sentences, may be based on the values of efficiency and speed in communication. That is, from an ideological perspective, the new conventions noted above move away from rules established within the traditional social institutional framework to those inscribed in the purposive-rational institutional framework.

But the wider significance of the restructuring of conventions (almost without conscious volition of users) is that it reflects and affects more than just the mode and style of the communication. It also reflects and affects how we interact with the technology, how we interact with others, and thus how the technology affects our lives. Perhaps email, like the telephone, does not fundamentally transform conversation (Miller, "This" 284–285; Spooner and Yancey). But as Miller suggests, "technology changes the constraints and thus the rhetorical situation; the technology does potentiate 'new relationships' between authors and readers," out of which genres may grow ("This 285); it at least affects beliefs about how we should live—ideology.

The ideology of email thus leads not only to the issue of genre, but points beyond it as well. We believe that the "social processes" Yates and Orlikowski discuss as the basis of genres of organizational communication are themselves at least partly determined by the ideology of technology, and that "the reciprocal interaction between institutionalized practices and individual human actions" (299) are conditioned if not determined by this ideology, reproduced and reenacted by practioners and students (Herndl); these social processes need to be comprehended by Habermas's critical theory (Blyer). If "genres can be viewed as social institutions that both shape and are shaped by individuals' communicative action" (300), then they exist like organizations themselves within larger cultural ideologies that find expression in the mega- and ill-defined genres of social, political, and economic forms and organizations of existence.

THE PURPOSIVE-RATIONAL GENRE OF EXISTENCE

Within the purposive-rational system in which email exists, users learn software and skills necessary to increase efficiency, speed, and productivity, not necessarily to maintain the traditional norms of an institutional framework. Relations—both personal and business—become focused on the technical production of communication. Fig. 3.1 on page 79, "Schematic Representation of the Evolution of Institutional Frameworks," therefore applies not only to social and corporate organizations, but to individuals as well. The translation of Habermas's chart in table 3.2 below summarizes the purposive-rational ideology applied to email in ways that raise questions about professional and personal life for practitioners and students to consider.

For example, based on Habermas's category of "type of definition," practitioners and students, like emailoholics, expect an immediate response to their email, thus "enslaving" the recipient who is obligated to respond immediately or risk violating the behavioral expectations established by the characteristics of "conditional predictions" and "conditional imperatives" of the purposive-rational system. The characteristics of "context free language" described by Habermas can illuminate the problem of students using Netspeak inappropriately in writing, making Baron's advice that "[we] must decide whether to employ conventions of informal speech . . . or assumptions about more formal writing" (410) more difficult to implement or teach. Likewise, the advice on making the content of email messages clear by using a "direct, concise style" (Benjamin; Bly) does not really address implications of Habermas's ideological category "functions of action type," dealing with means-ends relations: while advice about concise style does not provide students adequate rhetorical strategies to improve their writing, it does suffice for communication in the purposive-rational system based on technical rules. The means-end relations of production has become a foundation of education (see Watkins).

CONCLUSION

Email use increases productivity, which is evident in studies that show the use of email technology has increased the output of individual laborers and thus in turn the productivity of businesses. Email saves the average employee 326 hours each year, which results in a 15 percent–20 percent increase in productivity for each worker (Ferris 2000a). Even when businesses subtract the time workers spent dealing with nonessential email, such as spam, they still see a productivity gain of $9,000.00 per employee per year because of email (Ferris 2000a). However, while it is apparently not widespread, some organizations are limiting or banning email among their employees because they believe email is preventing necessary face-to-face communication (Farrell; Best; CNN.com).

Some studies have shown that regular email users speak to fewer friends, immediate family, or neighbors face-to-face during the week and have weaker ties with their communication partners than do people who do not use email (Kanfer). Many researchers have stated that while people may notice some of the problematic changes occurring because of email use, they are frequently unaware of the implications of those problems. People have an increasing dependency on email, both psychologically and logistically to complete their work (Hallewell). Many email users state that it is their preferred way to communicate and that they would be unable to complete their

TABLE 3.2
Habermas's Translated: Systems of Purposive-Rational (Instrumental and Strategic) Actions in Email

Habermas's Categories	*Habermas's Characteristics*	*Definition of Characteristics*	*Examples of Characteristics*
action-orienting rules *(What governs actions?)*	technical rules	Rules that are inherent in the technology	Format of email header (To/From...) Text formatting
level of definition *(What type of language is used?)*	context-free language	Language that is free of traditional "social" context	Lack of recognition of social/power hierarchies in salutations in email messages
type of definition *(What types of behavioral expectations occur?)*	conditional predictions conditional imperatives	If I take this action then this condition will be the result If condition exists then I will perform this action	If I send the message via email it will get there instantly. If I get this message immediately, then I will answer it immediately
mechanisms of acquisition *(How are behaviors acquired?)*	learning of skills and qualifications	Learning how to perform tasks necessary for production work to occur	Learn to use and understand the equipment/ software
function of action type *(What is the purpose of actions?)*	problem-solving (goal attainment, defined in means-end relationship)	Actions geared toward achieving objectives	Email to request action or accomplish other work-related task
sanctions against violation of rules *(What happens when rules are violated?)*	inefficacy: failure is reality	When rules are violated, the objective (end) is not attained.	Message not sent/received/acted on Message unintelligible (gibberish)
"rationalization" *(What is the reason for working within the system?)*	growth of productive forces; extension of power of technical control	Focus on increasing production through technology and adherence to technical rules	Amount of time spent on email Checking email on vacation/after hours/continually... addiction increase in number of communications able to "handle"

work if the email system were down and would be upset about the loss of email capabilities (Lamb and Peek; Collin).

Email users generally communicate more than nonusers of email; yet, they overestimate the amount of email they actually send and receive, which may relate to feelings of "communication overload" and be a factor in email being one of the top ten stresses in the work environment (Kanfer; Hallewell). The expanding technical control of email on the job and in the home affects the way people live by increasing the expectations of the amount of work required and the amount of time people must spend working. Most people, if they think about the increase in work at all, probably see it as just an incidental effect of email use, and not part of the essential nature of email technology. However, this analysis reveals that increased productivity is not a coincidental extension of email technology, but the result of it.

The phantom machine is purposive-rational ideology, which permeates every dimension of our being.

Notes

1. We note here that the division between work and leisure is not only socially constructed, but in communication media already virtual and thus easily crossed with symbols. We also note that on email, as on the Internet generally, work and recreation not only refer to the discourse of work and recreation, but to actual activities themselves (see Moskow and Katz).

2. Strate et al. also discuss the breakdown between private and public spheres in cyberspace (1–26).

3. Cf Geise's discussion of the "National Information Infrastructure," where he finds the social motive of users themselves as a countertrend, making a distinction between business and social uses of the network (152–158).

4. However, there are some possible technological, economic, and cultural limitations to this freedom; see Spooner and Yancey; Thorton; Dance.

5. For an ironically humorous discussion of how to conquer an email addiction, see *Are You Addicted to Email?* www.newbiesnet.com/addictedtoemail.htm.

6. While some email conventions seemingly are becoming standardized, Baron states, "pronouncements on email usage . . . sometimes reflect personal taste more than established linguistic conventions" (Baron 405). For example, "Virginia Shea, author of *Netiquette,* acknowledges that she made up many parts of her book as she went along" (Baron 405).

7. Although email has moved away from noting social distinctions in salutations, such social distinctions may exist, and as Lee notes, are evident in domains and subdomains that serve as indicators of hierarchy: "The shorter one's address, the higher one's status (Lohr 1994), perhaps because a longer address provides more information about a sender, thereby defining the person more by function and membership than by personal traits" (314). Lee points to the identifiers of countries and of organizations (for example, university and governmental organizations vs. com service providers) as indicators of status that contain not only technical but social informa-

tion. For our purposes, these URLs flag users and define the relations of writers and readers in terms of the technical rules governing URLs.

8. We note an interesting anecdote: After firing composer and friend Bernard Herrmann during the making of *Torn Curtain*, Alfred Hitchcock never spoke to him again; but he did email him (Mankiewicz).

9. The word *Hello* first came into use as a verbal greeting when the telephone was a new technology and its inventors were trying to establish a way of initiating a telephone conversation. Alexander Graham Bell recommended the word *Ahoy*, which was a call to hail ships, and Thomas Edison recommended *Halloo*, which was a traditional hunting call to the hounds, but eventually changed it to *Hello*, which subsequently became a face-to-face and telephone greeting in the United States (Baron 408).

WORKS CITED

Angell, David, and Brent Heslop. *The Elements of Email Style: Communicate Effectively via Electronic Mail.* Reading, MA: Addison-Wesley, 1994

Anonymous of Bologna. "The Principles of Letter Writing." Trans. James J. Murphy. *Three Medieval Rhetorical Arts.* Ed. James J. Murphy. Berkley: U of California P, 1971. 1–25.

Bakhtin, Mikhail. "The Problem of Speech Genres." *The Rhetorical Tradition.* Ed. Patricia Bizzell and Bruce Herzberg. Boston: Bedford, 1990. 944–964.

Baron, Naomi S. "Who Sets Email Style? Prescriptivism, Coping Strategies, and Democratizing Communication Access." *The Information Society* 18 (2002): 403–413.

Baugh, L. Sue, Maridell Fryar, and David A. Thomas. *How to Write First-Class Business Correspondence.* Lincolnwood, IL: NTC Publishing Group, 1995.

Benjamin, Susan. *Words at Work: Business Writing in Half the Time with Twice the Power.* Reading, MA: Addison–Wesley, 1997.

Berlin, James. "Rhetoric and Ideology in the Writing Classroom." *College English* 50 (1988): 770–777.

Best, Jo. "Phones 4U Bans Staff from Email." ZDNet UK. http://news.zdnet.co.uk/communications/0,39020336,39116502,00.htm. 2003.

Bly, Robert W. *The Encyclopedia of Business Letters, Fax Memos, and Email.* Franklin Lakes, NJ: Career Press, 1999.

Blyler, Nancy Roundy. "Habermas, Empowerment, and Professional Discourse." *Technical Communication Quarterly* 3.2 (1994): 125-145.

———. "Research as Ideology in Professional Communication." *Technical Communication Quarterly* 4.3 (1995): 285-313.

Bolter, Jay David. "Virtual Reality and the Redefinition of Self." In Strate et al. 123–137.

Boshier, Roger. "Socio-psychological Factors in Electronic Networking." *International Journal of Lifelong Education* 9.1 (1990): 46–64.

Buber, Martin. *I and Thou*. Trans. Walter Kaufmann. New York: Charles Scribner's Sons, 1970.

CBSNews. "Addicted to Email?" *CBSNews.com.* http://www.cbsnews. com/stories/2002/08/12/earlyshow/contributor/reginalewis/main518307.shtml. 2002.

CNN.com. "Firm Bans Email at Work." www.cnn.com/2003/TECH/internet/09/19/email.ban/. 2003.

Collin, Simon. "Integrating Email: From the Intranet to the Internet." Boston: Digital Press, 1999.

Dance, Frank E.X. "The Digital Divide." In Strate et al. 171–182.

Dobrian, Joseph. *Business Writing Skills: A Take-Charge Assistant Book*. New York: AMACOM, 1998.

Dubrovsky, V., S. Kiesler, and B. N. Sethna. "The Equalisation Phenomenon." *Human-Computer Interaction* 6 (1991): 119–146.

Ellul, Jacques. *The Technological Society*. New York: Alfred A. Knopf, 1976.

Farrell, Nick. "Council Bans Email." vunet.com. www.vnunet.com/news/1133447. 2002.

Ferris Research. "Email Boosts Employee Productivity." *NUA.com.* www.groupweb. com/cgi-bin/search/groupweb.cgi?ID=998113370. 2000a.

———. "Summary: How email Affects Productivity." *NUA.com.* http://www.ferris. com/rep/20000202/SM.html. 2000b.

———. "Volume of Corporate email Still Growing." *NUA.com.* http://www.nua. com/surveys/index.cgi?f=VS&art_id=905357026&rel=true. 2001.

Festa, Paul. "Email Has Come a Long Way in 30 Years." *CNET News.com.* http://news.com.com/2100-1023-274170.html?legacy=cnet. 2001.

Finholt, Tom, and Lee Sproull. "Electronic Groups at Work." *Organization Science* 1 (1990): 41–64.

Foucault, Michel. *The Archeology of Knowledge and the Discourse on Language*. New York: Pantheon Books, 1972.

Geisler, Cheryl, and Annis Golden. "Mobile Technologies at the Boundary of Work and Life." www.rpi.edu/~geislc/Manuscripts/Mobile.pdf. In progress.

Geise, Mark. "From ARPAnet to the Internet: A Cultural Clash and Its Implications in Framing the Debate on the Information Superhighway." In Strate et al. 141–159.

Graham, John R. "It's Changed the Way We Think and Behave: Who Do We Thank (and Curse) for Email?" *The Sales Trainer* 22.2. http://www.nbmda.org/what/trainer/Trainer_July_Aug02.pdf. 2002.

Gumpert, Gary, and Susan J. Drucker. "From Locomotion to Telecommunication, or Paths of Safety, Streets of Gore." In Strate et al. 29–46.

Hallewell, Bob. "Softening the Organisational Impact of Email." *Training Journal.* www.trainingjournal.co.uk/articles/hallewell.htm. 2002.

Habermas, Jürgen. *The Structural Transformation of the Public Sphere: An Inquiry into a Category of Bourgeois Society.* Trans. Thomas Burger. Cambridge, MA: MIT, 1989.

———. "Technology and Science as 'Ideology.'" *Toward a Rational Society: Student Protest, Science, and Politics.* Boston: Beacon Press, 1970. 81–122.

———. *The Theory of Communicative Action Volume 2: Lifeworld and System: A Critique of Functionalist Reason.* Trans. Thomas McCarthy. Boston: Beacon Press. 1987.

Hafner, Katie, and Matthew Lyon. "Talking Headers." *The Washington Post Magazine.* http://www.olografix.org/gubi/estate/libri/wizards/email.html. 1996.

Hardy, Ian. "The Evolution of ARPANET Email." http://www.ifla.org/documents/Internet/hari1.txt. 1996.

Herndl, Carl G. "Teaching Discourse and Reproducing Culture: A Critique of Research and Pedagogy in Professional and Non-Academic Writing." *College Composition and Communication* 44.3. (1993): 349–363.

Holdstein, Deborah. "Power, Genre, and Technology." *College Composition and Communication* 47.2 (1996): 279–284.

Howe, Peter J. "'Peanuts!' 'Popcorn!' 'WiFi!.'" *Boston Globe.* 13 July 2004: D1.

Huff, Chuck, and R. King. "An Experiment in Electronic Collaboration." *Interacting by Computer: Effects on Small Group Style and Structure.* Symposium conducted at the meeting of the American Psychological Association. Atlanta. 1988.

IBM. "Looking at the Real Impact of IBM Lotus Notes and Domino 6." http://www.lotus.com/news/news.nsf/link/aidantroy. 2002.

Inkster, Robert P. and Judith M. Kilborn. *The Writing of Business.* Boston: Allyn and Bacon, 1999.

<julichny>. "word!" Email correspondence former student. 4 April 2002.

Kanfer, Alaina. "It's a Thin World: The Association Between Email Use and Patterns of Communication and Relationships." Urbana-Champaign, IL: National Center for Supercomputing Applications. http://archive.ncsa.uiuc.edu/edu/trg/info_society.html. 1999.

Katz, Steven B. "Aristotle's Rhetoric, Hitler's Program, and the Ideological Problem of Praxis, Power, and Professional Discourse." Special issue on Power and

Professional Discourse, *Journal of Business and Technical Communication* 7.1 (Jan. 1993): 37–62.

———. “Ghosts of Technology.” *Voices: Journal of the American Academy of Psychotherapy* 22.3 (1986): 34.

Katz, Susan. *The Dynamics of Writing Review: Opportunities for Growth and Change in the Workplace.* Stamford, CT: Ablex Publishing Corporation, 1998.

Koku, Emmanuel, Nancy Nazer, and Barry Wellman. Netting Scholars: Online and Offline. *American Behavioral Scientist.* 43.10 (2000): 1681–1704.

Lamb, Linda, and Jerry Peek. *What You Need to Know: Using Email Effectively.* Sebastopol, CA: O'Reilly & Associates, Inc., 1995.

Lee, Judith Yaross. “Charting the Codes of Cyberspace: A Rhetoric of Electronic Mail.” In Strate et al., 2003.

Locker, Kitty O., and Stephen Kyo Kaczmarek. *Business Communication: Building Critical Skills.* Boston: McGraw-Hill. 2001. 148-165; 255–267.

Lohr, S. “Can Email Cachet=jpmorgan@park.ave?.” *New York Times.* 1994. In Lee. 314.

MacCormac, Earl R. “Men and Machines: The Computational Metaphor.” *Technology in Science* 6 (1984): 207–216.

Mankiewicz, Ben. Closing comments on *Torn Curtain.* Turner Classic Movies, 15 May 2004, 1:30–4:00 A.M.

Marx, Leo. *The Machine in the Garden: Technology and the Pastoral Ideal in America.* New York: Oxford UP, 1964.

McCormick, N. B. and McCormick, J. W. “Computer Friends and Foes: Content of Undergraduates' Electronic Mail.” *Computers in Human Behavior* 8 (1992):1 379–405.

Microsoft. “Hotmail Turns Three: Still the Leader After All These Years.” http://www.eu.microsoft.com/presspass/features/1999/07-12hotmail.asp. 1999a.

———. “Microsoft Acquires Hotmail.” http://www.eu.microsoft.com/PressPass/press/1997/dec97/Hotmlpr.asp. 1997.

———. “MSN Hotmail Announces Major Redesign.” http://www.eu.microsoft.com/presspass/press/1999/Jul99/Redesignpr.asp. 1999b.

———. “MSN Hotmail Continues to Grow Faster Than Any Media Company in History.” http://www.eu.microsoft.com/presspass/features/1999/02-08hotmail.asp. 1999c.

———. “MSN Hotmail Extra Storage Fact Sheet.” http://www.microsoft.com/presspass/newsroom/msn/factsheet/hotmail.asp. 2004.

———. “Next Generation MSN Hotmail.” (Flash Movie). http://newsletter.webdesign4india.com/index.php?p=49. Originally on www.hotmail.com.2003a.

———. "Outlook 2002 Product Guide (MPG)." http://www.microsoft.com/office/outlook/evaluation/guide.asp. 2001.

———. "Record Numbers of Consumers Are Reaching for MSN Hotmail and MSN Messenger as New Version of Hotmail Debuts." http://www.microsoft.com/presspass/press/2003/dec03/12-03SmartScreen2003PR.asp. 2003b.

Miller, Carolyn R. "Genre as Social Action." *Quarterly Journal of Speech* 70 (1984): 151–67.

———. "The Rhetoric of Decision Science, or Herbert A. Simon Says." *The Rhetorical Turn: Invention and Persuasion in the Conduct of Inquiry.* Ed. H. W. Simons. Chicago: U of Chicago P, 1990. 62–184.

———. "Technology as a Form of Consciousness: A Study of Contemporary Ethos." *Central States Speech Journal* 29 (1978): 228–236.

———. "This Is Not an Essay." *College Composition and Communication* 47.2 (1996): 284–288.

Moskow, Michal Anne, and Steven B. Katz. "Re-Building Life in Cyberspace: Ethnography of Rhetoric in Post-Holocaust Discussion Lists in Germany and United States." *Judaism and Composition.* NJ: Hampton Press, in press.

<mrrandal>. "Re: Feedback." Email correspondence from former student. 25 November 2003.

Nchor, Joseph. "Email and Productivity." *Groupweb.com.* http://www.emailtoday.com/emailtoday/pr/email_productivity.htm. 2001.

Orlikowski, Wanda, and Joanne A. Yates. "Genre Repertoire: The Structuring of Communicative Practices in Organizations." *Administrative Science Quarterly* 39 (1994): 541–574.

Pratt, Laurie A. "Impression Management in Organization Email Communication." http://www.public.asu.edu/~corman/scaorgcomm/pratt.htm. 1996.

Richardson, Malcolm, and Sarah Liggett. "Power Relations, Technical Writing Theory, and Workplace Writing." *Journal of Business and Technical Communication* 7.1 (1993): 112–137.

Rikes, Jeff. "The Productivity Frontier." http://www.microsoft.com/mscorp/execmail/2003/10-13productivity.asp. 2003.

Schultz, Heidi. *The Elements of Electronic Communication.* Boston: Allyn and Bacon. 1999.

Selwyn, Neil, and Kate Robson. "Using Email as a Research Tool." *Social Research Update.* http://www.soc.surrey.ac.uk/sru/SRU21.html. 1998.

Shea, Virginia. *Netiquette.* San Francisco: Albion Books. 1994.

Spooner, Michael, and Kathleen Yancey. "Postings on a Genre of email." *College Composition and Communication* 47.2 (1996): 252–278.

Sproull, Lee, and Sara Kiesler. "Connections." *New Ways of Working in the Networked Organization.* Cambridge, MA: MIT P, 1991.

Strate, Lance, Ron L. Jacobson, Stephanie Gibson. *Communication and Cyberspace: Social Interaction in an Electronic Environment.* 2nd ed. Cresskill, NJ: Hampton Press, 2003.

Terminello, Verna, and Marcia G. Reed. *Email:1 Communicate Effectively.* Upper Saddle River, NJ: Prentice Hall, 2003.

Thornton, Alinta. "Does Internet Create Democracy?" Oct. 2002. http://www.zip.com.au/~athornto. Accessed: 3 March 2004. Original article published in *Ecquid novi* 22.2 (2001).

Tschabitscher, Heinz. "Email Addiction (Emailoholism)." *About.com.* http://email.about.com/library/weekly/aa120897a.htm.

Weil, Debbie. "Email Addiction and the B2B Email Marketer." *ClickZ Experts.* http://www.clickz.com/experts/em_mkt/b2b_em_mkt/article.php/838401. 2001a.

———. "Finding a Cure for Email Addiction." *ClickZ Experts.* http://www.clickz.com/experts/em_mkt/b2b_em_mkt/article.php/839041. 2001b.

Yates, Joanne, and Wanda J. Orlikowski. "Genres of Organizational Communication: A Structurational Approach to Studying Communication and Media." *Academy of Management Review* 17.2 (1992): 299–326.

Zucchermaglio, Cristina, and Alessandra Talamo. "The Development of a Virtual Community of Practices Using Electronic Mail and Communicative Genres." *Journal of Business and Technical Communication* 17.3 (2003): 259–228.

Acknowledgments

We would like to acknowledge our colleagues—Carolyn Miller, David Reider, and Susan Katz—for their advice and guidance on the prototype of this article. Their comments were invaluable in helping us think through our arguments and reach higher ground.

Research

Numerous scholars have critiqued views of technical communication grounded in scientific positivism. In her groundbreaking essay "An Approach for Applying Cultural Study Theory to Technical Writing Research," Bernadette Longo takes on the more difficult task of challenging (by extending) social constructionist approaches of studying technical communication. Traditional social constructionist approaches limit their notion of cultural context to the dynamics within an organization, Longo explains, and stop short of critiquing the political functions and effects of technical communication. As an alternative, Longo offers a poststructuralist, Foucaultian cultural studies approach that situates technical communication in broader but often invisible "social, economic, political, and institutional" relations (125; 67 in original), that critiques technical communication's roles in power-laden struggles over knowledge legitimation (120; 61 in original), and that uncovers "silences, absences, and exclusions" still present in dominant discursive practices (126; 68 in original). Further, Longo challenges researchers to be attentive to how they constitute and limit knowledge through their methods (127; 68 in original). In her thorough discussion of this theoretical framework, Longo explains how a cultural studies method can be flexible but still coherent, rigorous, and valid. Its validity, she argues, should be measured in part by its effects, by what it teaches us about technical communication and by how well it opens up "further discourse as potential objects of future research" (122; 63–64 in original). By this definition, Longo's own research is certainly valid, as it has prompted a range of responses in this collection and elsewhere.

In her chapter "The Rhetorical Work of Institutions," Elizabeth C. Britt answers Longo's call to expand the focus of technical communication beyond discrete organizations to networks of institutions (e.g., law and medicine), though she views her project as partly building on the organization-focused work of others in the field. Britt draws on social and cultural theory from sociology and anthropology to conceptualize institutional practices as taken for granted but also constituted and legitimated through rhetoric, which makes them more amenable to critique. Like Longo, Britt is interested in critiquing institutional power dynamics—in this case, of insurance—that shape and are shaped by technical communication. Through a brilliant rereading of Catherine Schryer's study of letter writing practices in an insurance company, Britt explains how narrow critique can actually be used to reinforce hegemonic power relations, and she shows what a cultural studies-based institutional critique could add. In rhetorically producing and responding to risk, Britt shows, the institutional practices of insurance maintain profits by creating tension between collective and individual responsibility and by furthering the cultural trend toward individual risk-taking (143).

Jeffrey T. Grabill echoes Longo's critique of narrow, descriptive, and instrumentalist technical communication research methods, but he also usefully critiques some versions of cultural studies for privileging critical, academic reading over civic action and social change. Grabill thus argues for redirecting both traditions. Citing the work of anthropologists and critical rhetoricians, he calls for the rhetorical study of technical/professional communication as cultural and institutional labor in the "social factory" of everyday life (157). Importantly, such an approach would treat research *practices* as mechanisms for rhetorical agency (162). Beyond acknowledging the ideological and formal, procedural dimensions of methodology, Grabill seeks to develop specific research practices—defined as tactical, dynamic actions for constructing relations with other stakeholders in the study—that include initiation, access, participation, "studying up," local politics, communication, and sustainability (161). In a discussion of his own community-based study of environmental risk communication, Grabill illustrates how his practices for gaining the trust of community partners (and thereby creating access) and working with them to develop alternative means of risk communication created a space for activist agency for the researchers and for technical communication. "Research, then, produces cultural and it produces its own possibilities for change," Grabill concludes, "More important, critical research in technical and professional writing helps others be productive . . . in the institutional spaces that shape their lives (and that we hope they, in turn, can shape)" (167). While Longo and Britt open up the space for cultural, institutional critique and intervention, Grabill demonstrates how to operationalize these moves.

In this section's final chapter, Beverly Sauer joins Grabill in illustrating how technical communicators and researchers can both learn from and inform sociocultural theory (in her case, from sociology). Sauer's rich study of regulatory mining standards as "living documents" that are "continually revised and updated in response to new information, new technologies, and changing social and institutional practices" combines a sociocultural critique of power relations at the institutional level with a more nuanced rhetorical analysis of regulatory documents' functions in specific, dynamic, and hazardous local contexts. Sauer's study reveals important tensions between broad regulatory standards and the embodied, situated experiential knowledge from those on the front lines of risk decision-making; between a postpositivist uncertainty about science and the need for positivist notions to preserve worker health and safety; and between the need to archive the potentially useful situated knowledges preserved in outdated regulatory (print and online) documents and the need the expunge these documents to avoid misinformation or confusion. Sauer ultimately argues that cultural theory can help technical communicators negotiate these tensions by posing "questions about the nature of documentation practices and their relation to cultural history, power, and material, embodied experience" (175); at the same time, technical communication practitioners and researchers must be critical of our assumptions about the "representational character of language" and the effects of regulatory power (189).

As this overview suggests, these four chapters do more than simply impose cultural studies onto our field's dominant research methods. Instead, they self-reflexively develop hybrid approaches that move both technical communication and cultural studies in productive directions, directions that lead to a deeper, more critical understanding of communicative practices and the space to ethically intervene in them.

Chapter Four

An Approach for Applying Cultural Study Theory to Technical Writing Research

Bernadette Longo

> Merely qualitative knowledge leads to the grossest errors of judgment, and is of that kind of little learning which is a dangerous thing.... All engineering problems are purely quantitative from the beginning to the end, and so are all other problems, whether material, or moral, or financial, or commercial, or social, or political, or religious. All judgments passed on such problems, therefore, must be quantitative judgments. How poorly prepared to pass such judgments are those whose knowledge is qualitative only!
>
> —John Butler Johnson, 1928

> Once every text had to take account of the theologians and the authorities, so today every text is written against the back-drop of science and the authority of truth with a small t.... And without this truth the distinctions between science and art, between literature and philosophy, between stories and truths, begin to look precarious.
>
> —Hilary Lawson, 1989

Good technical writing is so clear that it is invisible. Yet technical writing is the mechanism that controls scientific systems, thereby organizing the operations of modern institutions and the people within them. The invisibility of technical writing attests to its efficiency as a control mechanism because it works to shape our actions without displaying its methods for ready analysis. When technical writing is made visible for study, it is often characterized as a simple collaborative effort in which writers mediate technology for users. Yet it can also be seen as a mundane discourse practice working to legitimate some types of knowledge while marginalizing other possible knowledges.

This chapter originally appeared in *Technical Communication Quarterly* 7.1 (1998): 53–74. Reprinted with permission of Lawrence Erlbaum Associates, Inc. and the Association of Teachers of Technical Writing.

These questions of knowledge legitimation within cultural contexts extending beyond the walls of one organization cannot be addressed well through social constructionist research designs. They can be addressed, however, through cultural study.

During the course of the twentieth century, our concepts of technical writing have developed in tandem with other elements of our culture. Like any other aspect of our culture, therefore, technical writing can be fruitfully explored through a cultural studies research framework. This article will explore how cultural studies are helping to illuminate technical writing issues left unseen in more scientifically modeled research studies—issues such as effects of institutional relationships and expanded notions of culture on technical writing practices, as well as the cultural power of knowledge produced through technical writing. This article will also discuss some implications of using a speculative research model in institutions where nonscientific knowledge is often seen as nonlegitimate. Finally, this article will offer some guidelines for designing a cultural study of technical writing, focusing on how to limit the study and providing some theoretical reasoning for this delimiting process.

PUTTING TECHNICAL WRITING PRACTICES IN CULTURAL CONTEXTS

Researchers in technical writing have begun to explore how technical writing is involved within situated institutional relationships of knowledge and power—how some types of knowledge are legitimated through technical writing practice, while other possible knowledges (such as knowledge gained through speculation, emotion, or intuition) are subjugated or excluded as marginal. In order to carry out such a critical analysis of technical writing, institutions where technical writing is practiced need to be reconstructed as cultural agents that are not necessarily bounded by any one organization's walls. Vincent Leitch described how institutions as cultural agents influence discourse:

> Through various discursive and technical means, institutions constitute and disseminate systems of rules, conventions, and practices that condition the creation, circulation, and use of resources, information, knowledge, and belief. Institutions include, therefore, both material forms and mechanisms of production, distribution and consumption and ideological norms and protocols shaping the reception, comprehension, and application of discourse.... Institutions often enable things to function, inaugurate new modes

> of knowledge, initiate productive associations, offer assistance and support, provide useful information, create helpful social ties, simplify large-scale problems, protect the vulnerable, and enrich the community. (127–128)

Because institutions are cultural agents that affect discourse practices, a study of technical writing as situated in systems of knowledge and power is incomplete if the idea of culture is limited within one organization. Recognition of an organization's participation in cultural contexts enables a critical study that can illuminate assumptions about the inevitable roles of technical writing in our culture at given historical moments—roles such as information conveyor or technology mediator.

A view of culture that is limited within the walls of one organization does not allow researchers to question assumptions about technical writing practices because those practices are not placed in relationship to influences outside the organization under study. This limited view of culture is tacitly adopted in traditional social constructionist research designs, however, as illustrated by Stephen Doheny-Farina's exploration of technology transfer *Rhetoric, Innovation, Technology.* In this study, technical writing was seen as participating in a "series of personal constructions and reconstructions of knowledge, expertise, and technologies by the participants attempting to adapt technological innovations for social uses" (ix). Although Doheny-Farina placed technical writing practice in a social setting, he assumed that problems in technology transfer have more to do with the quality of the technical texts produced by one organization than with economic, political, or social pressures or conflicts affecting a situated writing practice. Doheny-Farina could describe how a text shaped practice, but did not question why the text includes the information that it did and not other information that would be equally possible to include—why the text legitimated some kinds of knowledge and not others. And further, what systems of power (the cultural context of the organization under study) did the knowledge legitimated in the text uphold and what other possible systems of power did the text make impossible? These research questions could not be examined using a social constructionist research design that limited the idea of culture within one organization. To answer questions about knowledge legitimation would require an assumption that the culture within the organization under study was in tension with a wider culture that included related institutions.

Similarly, when Doheny-Farina analyzed the Paradis, Dobrin, and Miller study of "Writing at Exxon ITD," he found that their writing practice "becomes part of the process of developing an organizational identity; it becomes part of the process of group membership" (*Rhetoric, Innovation* 28). Social influences which form the culture under investigation in this analysis of the Exxon study were limited to the interior of the Exxon Corporation. The "social" in "social constructionist" here extended to a group of people within one organization. Because of this limited view of culture, Doheny-Farina was not able to explore how Exxon's texts interacted with texts from competing companies, government agencies, and the legal system, for example.

The limitations of confining culture within the walls of one organization were further illustrated in Doheny-Farina's consideration of a 1991 article by Herndl, Fennell, and Miller in which these authors examined miscommunication and misunderstanding in the Three Mile Island and Challenger disasters. Doheny-Farina constructed the communication problems that contributed to the Challenger disaster as problems of technology mediation: "Issues of texts miscommunication and misunderstanding . . . involve the failures of texts to mediate technology to users" (*Rhetoric, Innovation* 28). This focus on the technical writer's role as mediator between technicians and users allowed Doheny-Farina to explore how technical writers upheld science and technology's dominant position within our culture. But it did not question what influences other than miscommunication might have contributed to the misunderstandings described in the Herndl et al. article. For example, the decision to go ahead with the Challenger launch was made despite evidence of previous O-ring erosion and a forecast of record cold weather at launch. Morton Thiokol engineer Roger Boisjoly argued for safety and against the launch at a meeting with NASA representatives the night before the Challenger was launched. He later suggested an extratextual factor creating pressure to launch: Morton Thiokol was in the process of negotiating a $1 billion contract with the U.S. government for space shuttle parts and the government was considering second-sourcing Morton Thiokol. NASA staff were intent on launching and, during the prelaunch meeting, asked Morton Thiokol management to "rethink" their recommendation not to launch. Morton Thiokol reversed its no-launch recommendation to acquiesce to NASA's wishes, despite Morton Thiokol engineers' warnings about unsafe launch conditions. More than a simple case of misunderstanding, discourse surrounding the decision to launch the Challenger was influenced by economic and political considerations that legitimated some knowledge (the managers' judgment and decision to launch) and marginalized other knowledge (the engineers' data and warnings). The knowledge legitimated in this case was clearly participating in systems of institutional power that extended far beyond Morton Thiokol's walls.

A limited view of culture is not unusual in technical and professional communication research. (See the following examples of social constructionist research which limit the idea of culture to one organization: Jennie Dautermann, "Negotiating Meaning in a Hospital Discourse Community"; Amy Devitt, "Intertextuality in Tax Accounting"; James Porter, "The Role of

Law, Policy, and Ethics in Corporate Composing"; Graham Smart, "Genre as Community Invention"; Gail Stygall, "Texts in Oral Context"; and Charlotte Thralls, "Rites and Ceremonials.") In an example of separating a government agency's discourse practices from their context of political influences, Susan Kleimann assumed that culture resides within one government organization (the General Accounting Office) in her study entitled "The Reciprocal Relationship of Workplace Culture and Review." Kleimann confined the idea of culture either to divisions within the General Accounting Office (GAO) or to the GAO as an isolated governmental entity and described how the organization's values influenced document review practices:

> Two major influences shape aspects of GAO's culture. First, GAO reports often result in changes to national policy, legislation, and funding; second, until recently, most GAO employees were educated as accountants, valuing minutiae and accuracy. Consequently, the agency has a cautious culture that demands maintaining detailed and extensive workpapers, referencing all facts to these workpapers, wanting both accuracy and objectivity and requiring an extensive review process. Moreover, because of the reports' potential impact, review is used to ensure that reports are correct from several perspectives. (58)

While Kleimann noted that GAO texts are influential in shaping national policy, legislation, and funding decisions, she confined the culture discussed in her study within the boundaries of the GAO. While she studied one of the major institutions in any culture (the government), she did not ask how national and international political tensions shaped GAO texts and how politics were in turn shaped by GAO texts.

The limited and conservative view of culture found in much research in technical and professional communication was articulated in Jack Selzer's description of intertextuality: "Indeed, 'context' or 'environment' or 'setting' or 'culture' might be understood as nothing more than a complex of language and texts, and individuals within an environment therefore might be understood as minds assimilated into its concepts and terminology" (172). Or further: "Readers and writers experience the flow of culture as a kind of collaboration among seen and unseen authors and texts and readers, all in the process of making sense. One job of the critic of culture—of the rhetorician, that is—is to uncover the various resonances inscribed in the tapestry of text and to account for their source, their intricacy, and their meaning" (179). While Selzer's view of culture acknowledged that cultural criticism involves itself with meaning, his view did not involve texts in tensions within situated relations of power and knowledge—where knowledge legitimation is contested by various interest groups and "making sense" means something different depending on your point of view. To view culture as a "kind of collaboration" works to sanitize what Walter Benjamin described as barbarism inherent in the spoils of war (for cultural legitimation):

> Whoever has emerged victorious participates to this day in the triumphal procession in which the present rulers step over those who are lying prostrate. According to traditional practice, the spoils are carried along in the procession. They are called cultural treasures... There is no document of civilization which is not at the same time a document of barbarism. And just as such a document is not free of barbarism, barbarism taints also the manner in which it was transmitted from one owner to another. ("Theses" 256)

"Making sense" within a framework of contests for knowledge legitimation is not merely a "kind of collaboration." From a critical point of view, making sense for the victor is not making sense for the vanquished, who might ask why their knowledge was silenced.

In *The Differend*, Jean-François Lyotard described the silencing of nonlegitimated knowledge as a "wrong" suffered in "a case of conflict, between (at least) two parties, that cannot be equitably resolved for lack of a rule of judgment applicable to both arguments" (xi). Because there is no universal rule for equitable judgment, actions taken through discourse must privilege one way of knowing over other possible ways of knowing. Unlike a simple idea of consensus-based collaboration, Lyotard's theory of discourse production holds that power is distributed unevenly among possible ways of knowing. Basing his description on how phrases are linked in discourse, Lyotard found, "In the absence of a phrase regimen or of a genre of discourse that enjoys a universal authority to decide, does not the linkage (whichever one it is) necessarily wrong the regimens or genres whose possible phrases remain unactualized?" (xii). Using Lyotard's theory, discourse becomes a contest for legitimating knowledge and culture is more hegemonic than simply collaborative. Discourse becomes a struggle mediated by culture.

Struggles in discourse are contained within the discourse. For Michel Foucault, discourse holds histories of struggles for knowledge legitimation and the articulated discourse subsumes other discourses that were possible, but not articulated. In arguing for the study of culture through discourse analysis, Foucault described how the legitimated knowledges articulated in discourse embody historical struggles for their legitimation and conquest:

> In the two cases—in the case of the erudite as in that of the disqualified knowledges—with what in fact were these buried, subju-

> gated knowledges really concerned? They were concerned with a *historical knowledge of struggles*. In the specialized areas of erudition as in the disqualified, popular knowledge there lay the memory of hostile encounters which even up to this day have been confined to the margins of knowledge. (*Power/Knowledge* 83)

At the margins of knowledge we can find two types of delegitimated knowledges: erudite learning that may have been previously legitimated knowledge, but has been subsumed (or conquered) by other subsequently legitimated knowledges; and naive "know-how" (de Certeau, *Everyday Life* 65 ff.) that was previously legitimated as sufficient for carrying out everyday practices, but also has been subsumed by other subsequently legitimated knowledges (usually some sort of science or theory).

Because technical writing participates in institutional relationships, it works to organize knowledge through science and practice through theory. For example, because technical writing knowledge is made through institutions such as schools and publishing companies, naive or uneducated technical writing practices are organized through academic and economic systems, which tend to reproduce our culture's dominant scientific model. This organizing activity is found by Michel de Certeau to be a trend in Western culture since the time of Francis Bacon: "the sciences are the operational languages whose grammar and syntax form constructed, regulated, and thus writeable, systems; the arts are techniques that await an enlightened knowledge they currently lack" (*Everyday Life* 66). Because science forms the legitimated language of practice, it is a "writing that conquers" (*Writing of History* xxv) other practices based on naive "know-how":

> But at the same time that they acknowledge in these practices a kind of knowledge preceding that of the scientists, they have to release it from its "improper" language and invert into a "proper" discourse the erroneous expression of "marvels" that are already present in everyday ways of operating. Science will make princesses out of all these Cinderellas. The principle of an ethnological operation on practices is thus formulated: their social isolation calls for a sort of "education" which, through a linguistic inversion, introduces them into the field of scientific written language. (author's emphasis, *Everyday Life* 67)

If technical writing is the mediator between technology and what we have come to term *users*, technical writing practices work to conquer users' naive know-how and reformulate these uneducated practices into scientific discourse that can partake of the cultural power residing in scientific knowledge.

In so doing, technical writing participates in a writing that conquers naive knowledge by educating it into the technologies of scientific disciplines. Thus, technical writing participates in a system of knowledge and power within our culture (conceived as institutional relationships extending beyond the walls of any one organization). Technical writing also enables technology users to participate in our culture's knowledge/power system—a system that can be illuminated and analyzed using critical approaches to discourse practices. If, for example, technical writers' practices were studied within one organization (as in social constructionist research designs), the resulting analysis would not be able to show how the texts these writers produced were shaped by—and worked to shape—texts in related institutions, such as government regulations, competitors' work, economic policies, laws, and so on. Researchers can analyze these textual workings of our culture's knowledge/power system, however, using a cultural studies design.

CULTURAL STUDIES IN TECHNICAL WRITING

If social constructionist research in technical and professional communication cannot illuminate issues of knowledge and power, some recent cultural studies illustrate how these issues can be explored using an extended idea of culture. In a study based on poststructural theory, Richard Freed advocated widening our views of discourse communities beyond "company specific" boundaries to enable analysis of intercompany relationships (213, n. 8), thereby bringing corporate relationships into play and expanding culture beyond one organization. Freed's work also illustrated how using Lyotard's notions of grand narratives and *petit recits* as set out in *The Postmodern Condition* enables researchers to reconceptualize knowledge as historically situated: "Because the shape and tonality of knowledge vary by locale, and because for that locale the tone and temper of its knowledge rings and feels true, truths at one locale may be different from those held self-evident at other sites and from those held at different times at the same locale" (204). Freed's work further illustrated how using Baudrillard's notion of simulacra problematizes roles of texts, authors, and subjects within professional communication:

> [I]n a world of copies, of simulations without origins, originality means assembling copies in a new way, like recombinant DNA, by taking existing bits and bytes of text and recombining them. Thus not only is the author already written, by the prescribed roles that the organization or group authorizes him to play, but his materials for discourse are already inscribed, in the intertextual system that

> allows him to speak. This so-called death of the author, however, doesn't mean his demise, only that the scribe is always already circumscribed. (212)

Using critical theory allows Freed to explore how technical and professional communication work with situated knowledge that is profoundly shaped by contests for legitimation and which, in turn, shapes subsequent discourse and knowledge.

Other researchers have begun to consider how discourse practices participate in institutional relationships. For example, Bruce Herzberg advocated that composition researchers use Foucault's archaeological research approach "to analyze more closely the role of our institutions and disciplines in producing discourse, knowledge, and power" (80). He asked this question of the social constructionist research paradigm: "[W]hen the group agrees on standards for sufficient evidence or adequate organization or coherent argument, what is the source of its authority?" (79). This questioning of the basis of authority opens up discussions of how discourse participates in power/knowledge systems. In other words, what power sanctions the authority of the knowledge that is described in observed discourse practices? Ben and Marthalee Barton also asked this question of power and authority in exploring the practice of cartography. They found that maps as discourse practices were closely linked to institutional systems of power and knowledge: "Ultimately, the map in particular and, by implication, visual representations in general are seen as complicit with social-control mechanisms inextricably linked to power and authority" (53). These two examples of institutional critique illustrate how discourse can be seen as participating in systems of knowledge and power, exploring why some knowledge is articulated and legitimated while other possible knowledge is marginalized or left silent.

In another example of how critical theory can inform writing research, Lester Faigley explored how subject positions we constitute in composition classrooms rely on teachers' roles "as representatives of institutional authority" (*Fragments* 130). By applying Foucault's archaeology method, Faigley found a technology of confession in writing classrooms where personal narratives are seen as productive of "truths." This technology reproduces existing relations of knowledge and power between teachers and students, in which teachers dominate and students are dominated: "Such an assignment of authority through a teacher's claim to recognize truth is characteristic of Foucault's description of the modern exercise of power. Foucault writes that power is most effective when it is least visible" (*Fragments* 131). Using Foucault's archaeology enabled Faigley to discuss issues of institutional power and individual subject positions in the composition classroom because Foucault's work, and critical theory in general, provides a theoretical basis for recognizing institutional relationships and a vocabulary for discussing power/knowledge systems.

Recent cultural studies of technical writing and composition point to the fruitfulness of an approach based on Foucault's archaeological research methods and augmented by closely related lines of critical theory. Applied to technical writing, this approach can illuminate how struggles for knowledge legitimation taking place within technical writing practices are influenced by institutional, political, economic, and/or social relationships, pressures, and tensions within cultural contexts that transcend any one affiliated group. This type of cultural study can help to answer questions about why technical writing practices work to legitimate some types of knowledge while marginalizing other possible knowledges.

In order to look at struggles for knowledge legitimation that take place within technical writing, researchers can begin by asking Foucault's question, "How is it that one particular statement appeared rather than another?" (*Archaeology* 27). The statements that did appear in technical texts retell stories of the struggles, contradictions, and tensions within historic relations of knowledge and power. These statements also hold the silence of other statements that were possible but did not appear in technical texts at the particular time and place under study. By looking at statements that did appear and positing possible statements that did not appear, the genealogical researcher can construct what Foucault called a "systematic history of discourses" (*Birth* 14). Such a systematic history (or genealogy) of discourse asks questions about how one possible discourse was produced and legitimated as knowledge through technical writing, while other possible discourses were not produced and legitimated. This question of how one group's discourse became knowledge within a situated culture while another group's discourse was not seen as knowledge strikes at the heart of current discussions within technical and professional communication: multiculturalism (Bell, Dillon, and Becker; Kossek and Zonia; Nicholson et al.; Beverly Sauer; Michele Wender Zak), postmodernism (Ben and Marthalee Barton; Britt, Longo, and Woolever; Richard Freed), gender issues (Boiarsky et al.; Sam Dragga; Griffeth et al.; Mary Lay), conflict (Rebecca Burnett; Carl Herndl; McCarthy and Gerring), and ethics (Stephen Doheny-Farina; Steven Katz; James Porter). Therefore, issues such as multiculturalism, postmodernism, gender, conflict, and ethics within technical and professional communication can be analyzed using a cultural studies research design.

WHAT ABOUT OBJECTIVITY? WHAT ABOUT VALIDITY?

Adopting a cultural studies approach leaves a researcher open to criticism that the results of this type of speculative research are not objective and, therefore, are not valid. In other words, how are the results of cultural studies research to be understood in comparison to the results of scientifically modeled research? Choosing a coherent theoretical framework for a cultural study is crucial to understanding its results.

Cultural studies has been described by Cary Nelson, Paula Treichler, and Lawrence Grossberg as an "aggressively anti-disciplinary" activity drawing on many bodies of theory and having no distinct methodology (2). If this is the case, a researcher who sets out to do a cultural study must carefully construct a theoretical foundation to underpin her work, since there is no ready-made disciplinary or theoretical foundation for this type of study. The researcher can choose from many bodies of theory and/or methodology for constructing this foundation, such as anthropology, sociology, feminism, (neo-)Marxism, or poststructuralism (to list a few possibilities). In the process of choosing a body of theory, methodology, or both to work with and setting out a theoretical foundation for the work, the researcher will necessarily reflect on the nature of the study and his or her relationship to the object of inquiry. In this process, the researcher will develop an explicit rationale for choosing a body of theory to form a foundation for the cultural study. By considering technical writing practice and the research question within a context, the cultural researcher cannot hope to adopt a research stance outside the context of the object of inquiry. Technical writing must be studied within its context of institutional relationships, which includes the researcher as only one of many potentially "confounding" factors.

This relationship of the researcher to his or her object of cultural study bears closer scrutiny, since this relationship precludes claims of "objectivity," which are most often goals of modern academic study. In using this concept of "objectivity," I intend the term in a realist or positivist sense, in which it means something like this: observing the phenomenon without interference from (confounding) human factors. Other definitions of "objectivity" are possible. For one instance, pragmatists might define "objectivity" as sufficient intersubjective agreement for the phenomena to be viewed as "real" rather than as subjective. A postmodern view might see pragmatic "objectivity" as opposed to the realist "objectivity" and, therefore, participating in the modernist search for a metaphysical idea of "objectivity" through a progressive science or philosophy. In responding to these modernist notions, a postmodern view might define "objectivity" as impossible or irrelevant. (See Richard Rorty's "Science as Solidarity" and Hugh Tomlinson's "After Truth" for a discussion of pragmatic and postmodern views of objectivity.)

If objectivity is not possible, how is a cultural study to be considered "valid" by other academic researchers, especially those who conduct research in the quantitative model? How are readers to understand and evaluate the results of a cultural study?

The decentered, unstable author position resulting from self-reflexive poststructural research necessarily impacts traditional ideas of validity. Instead of asking, like Don Stacks and John Hocking, "whether a measurement technique...provides measures of what its user thinks it is measuring" (127), a poststructural view of validity would ask (following D. Gordon and Patti Lather) whether the research is an "'incitement to discourse'" (Lenzo 19)—that is, does the research prompt further discourse as potential objects of future research. This further discourse would indicate that researchers and other interested parties considered the findings relevant for continued conversations. In light of this self-reflexive and poststructural reconstruction of the idea of validity, a cultural study can be understood as situated—a questioning of representation and reality as they have been constructed at a certain place and time. This idea of validity also suggests that a cultural study will be partial and influenced by the researcher who undertakes the research. In other words, a poststructural cultural study can be "valid" according to a revised and enlarged notion of validity, but it will not be an "objective" study.

The assertion that a cultural study can achieve validity without objectivity flies in the face of centuries of scientific research—a radical position not to be adopted without understanding the scientific tradition against which it is positioned. The scope of this article certainly does not permit a discussion of the development of the Western scientific tradition to thoroughly trace the tension between science and speculation into which cultural studies must be placed. It is important to trace this tension, however incompletely, because it is an example of how our culture legitimates scientifically derived knowledge while marginalizing knowledge derived through speculation (such as practiced in cultural studies).

To understand cultural studies results in relation to results of more scientifically modeled research, I will briefly trace a contest for knowledge legitimation between experiment and speculation to the works of Francis Bacon, who advocated a natural philosophy based on experience and argued against a philosophy based on logical speculation in the scholastic tradition. Bacon is generally credited in histories of Western thought with articulating an argument for scientific knowledge based on methodical observation of natural phenomena, with the "systematic reform of natural philosophy" (Creighton viii), or with seeking "to reform the whole of human learning" (Zappen 61). Bacon can also be seen (from a poststructural point of view) as caught up in a contest for knowledge legitimation being waged between the dominant Scholastics and the upstart Reformationists. Francis Bacon in one sense sought to free "truth" from the hold of the Christian Church and a Scholastic tradition based on Aristotelian logic. Bacon's project for a public science promised to improve the human condition through a methodical, rational study of Nature. But in focusing on a public science and an improved human condition, Bacon necessarily addressed social and political concerns. Bacon's social project is concerned with reforming and securing the governance of the state as much as with improving the condition of its inhabitants. It is an art of managing people that Bacon sets out as much as it is a project for dominating Nature.

From our vantage point, we can see that Bacon's project for a modern, method-driven public science has been successful. This project, begun three centuries ago, overthrew the once-dominant scholastic tradition. In accomplishing this overthrow, the authority of modern science replaced the once-dominant logical speculation with the now-dominant scientific method. Scientists have accomplished this replacement by their appeal to utility and progress for the improvement of the human condition and they have, indeed, produced improvements in living conditions. Because of these improvements, people who deal in speculative or other nondominant ways of knowing agree to participate in the hegemony of the sciences.

Those of us living in the United States at the late-twentieth century take for granted the dominant place of science in our culture. We often assume that science gives us objective truths. We do not readily see scientific dominance as an outcome of contests for knowledge legitimation that came to a head in our culture some three hundred years ago—and are still waged today.

When scientific discourse is seen as producing objective truths, it is not seen as participating in culturally contextualized contests for dominance and power. In order to see how scientific and technological discourses participate in cultural contests, Stanley Aronowitz argued that these discourse practices must be placed in cultural contexts that extend beyond laboratories or one company's R&D department: "The point is...to situate science as a discourse within a larger system of social relations in which economic and political influences do not necessarily appear directly in the laboratory" (19). Since technical writing practices are so closely related to science, the same could be said about them: the point is to situate technical writing as a discourse within larger systems of cultural relations that do not necessarily appear directly wherever the practice takes place. These larger cultural relations may not appear directly because we assume many of them to be natural states of affairs, invisible relations enmeshed in intricate webs of institutional influences that appear to be inevitable.

One point of studying technical writing as a cultural practice is to make visible what seems invisible in technical writing, to view what seems inevitable as a product of culture. But since this approach to studying technical writing could potentially include everything in its situated cultural context, the researcher needs to limit the study by constructing a coherent theoretical framework for focusing the study and describing the object of inquiry in its institutional relationships.

DELIMITING THE OBJECT OF CULTURAL STUDY

If a cultural study can potentially include everything within a culture, how can a researcher construct a framework from which to study technical writing? If a cultural study cannot claim objectivity and relies on a test of validity that calls for the chosen methods to "incite discourse," what methodology can a researcher use to guide the construction of this framework?

Unlike quantitative research projects, which can rely on standard methods to guide the research design, cultural study remakes guidelines anew each time. Instead of relying extensively on standards, a cultural researcher relies on analysis of the situatedness of technical writing practices within a cultural context (which includes the researcher) to guide the research design. In the absence of scientific standards and methods, a cultural researcher relies on self-reflexive and nonscientific approaches to understanding the object of inquiry within a cultural context. Underlying this self-reflexive approach to poststructural cultural study is a philosophy that people know the world through language and that we construct realities through language. Because of this primacy of language in poststructural theory, cultural studies focus on discourse practices (such as technical writing) as objects of study.

In designing a cultural study of technical writing, a researcher can reflect on the following five (of many possible) themes: the object of the study is discourse, the object is studied in its cultural context, the object is studied as historically situated, the object is ordered by the researcher for the purposes of the study, and, therefore, the most important relationship in the study is between the object and the researcher. By reflecting on concepts such as these to design a cultural study, the researcher will be building a coherent framework of poststructural theory. The following brief discussions of these five themes provides a place to start thinking about their impact on technical writing research design.

The Object as Discourse

The concept of discourse as an object of inquiry is fundamental to work in cultural studies, wherein knowledge is viewed as constructed and legitimated through language practices. For this reason, technical writing is an especially apt object for cultural study. In *The Archaeology of Knowledge*, Michel Foucault focused on the concept of discourse as an object of inquiry in setting out his archaeology methodology, describing it as "a systematic description of discourse-objects" (140). The purpose of an archaeological study is to describe how rules of ordering statements both construct knowledge that appears in these statements and subjugate other possible knowledges that remain absent from the statements. Michel de Certeau in *Heterologies* and François Lyotard in *The Differend* further explored how nonlegitimated practices struggle with legitimated language practices to be heard within hegemonic social and political contexts. This consideration of the discourse object within complex cultural contexts is the basis of cultural studies. Since technical writing is clearly discursive, it can be readily constituted as an object of cultural study according to this theoretical concept.

Object within Cultural Context

Because technical writing is a practice that takes place within the context of dominant and subordinated knowledges (scientific and humanistic traditions, for example), constituting it as an object of cultural study can illuminate its role in complex social, economic, political, and institutional relations, which cannot be seen as well using quantitative or social constructionist methods. The study of cultural objects within their contexts is a practice of cultural studies which can be traced through early works in this field. Work done in the Birmingham Centre and reported by Stuart Hall et al. in *Culture, Media, Language* can be seen to reflect both the influence of Antonio Gramsci's writings in the *Prison Notebooks* and his notions of hegemony, combined with the Centre's earlier anthropological influences. From this early work, cultural studies has adopted anthropological notions that objects of inquiry can include everyday practices that were not seen as significant in the knowledge-making of some other disciplines. In *The Practice of Everyday Life*, for example, de Certeau clearly illustrated the importance of studying the contributions of mundane practice, explaining how people can accomplish nonlegitimated actions within a dominant culture. Technical writing can be seen as one of these mundane practices, whose role within a cultural context can be more completely described using a cultural studies research methodology in tandem with other more scientifically modeled methods.

Object within Historical Context

Because technical writing practices can be considered as being shaped by—and shaping—their cultural contexts, a cultural study of these practices will select significant moments in history to describe the relationships between technical writing and the surrounding culture. In effect, this type of study will often take the form of a descriptive history about the object of inquiry at a specific time and place. In *The Birth of the Clinic*, Foucault called for studying functional segments of discourse to create a systematic history that does not assume any remainder of meaning in the discourse. In other words, his purpose is to describe discourse-objects in their historical cultural contexts, without attempting interpretations of the objects. De Certeau in *The Writing of History* also argued that discourse should be studied in its historical aspects and as resulting from practices. He further saw the impulse that prompted the historical study as originating in current events. We may study cultural objects in their local, historical contexts, for example, but the impetus for the study exists in our present relations with the cultural objects. This concept tells us that technical writing can be constituted as an object of cultural study by placing it in a historical context and describing it within that context. The purpose of this type of study is not to interpret explicitly the role of the object within the context, but more to set out a history of that relationship of object and context, acknowledging that any discourse involves interpretation. The history a cultural study seeks to represent includes, as much as possible, subjugated knowledges as well as dominant ones.

Object as Ordered

In all the cultural studies mentioned so far, Walter Benjamin would have said that the one who undertook the study imposed an ordering on the object and its contexts, thereby including some ways of understanding the object and excluding others. He elaborated on this theme in "Theses on the Philosophy of History" when he argued that cultural objects are the spoils of the victor, but still hold within their silences the barbarism of the victory. This theme is reflected in many cultural studies' concern for difference, knowledge legitimation, and otherness. In *Power/Knowledge*, Foucault clearly acknowledged that some ways of knowing are subjugated, but are still present in language even in their ostensible absence. He called for us to create difference in discourse as an object of study, to uncover the subjugated knowledge, and then use the subjugated knowledge tactically in practice. In exploring the tactical nature of subjugated knowledge, Foucault described power/knowledge systems in military terms. In *The Differend*, Lyotard looked at this same type of inclusion/exclusion practice in terms of a legal system, describing a process whereby the linking of phrases, while necessary, is indeterminate as to what types of links can be made. In other words, phrases *must* be linked, but how they are linked is arbitrary, and many possibilities for linking exist. In choosing one way of linking, we necessarily exclude other ways, thereby silencing the knowledge that could be made through those other linkings. A cultural study of technical writing would explore those silences, absences, and exclusions still held within the dominant knowledge and discourse of that field's practices.

Relationship of Object with the One Who Studies

The person who researches technical writing practices necessarily orders them for the purposes of the study. Through this process, the researcher includes some possible ways of ordering and making knowledge and excludes others. In the research process, therefore, the most important encounter takes place between the object of study and the person who studies it. Technical writing is constituted in its encounter with the researcher, and it is limited by the researcher. Foucault in *Power/Knowledge* called for constituting the object of study as difference and discontinuity. For de Certeau in *Heterologies*, the aberrant can suggest an object for study. But whatever the researcher constitutes as an object of inquiry, remember Benjamin's observation that objects of cultural study, such as technical writing, are caught in a dialectic tension of chaos and order, present and past, rational and nonrational. A cultural researcher seeks to describe technical writing while recognizing his or her participation in these dialectical tensions that affect professional practices and the study.

A cultural study describes an antidisciplinary, situated, and personal encounter between a researcher and technical writing practices. When this study is firmly grounded in coherent theoretical rationales and employs a rigorous methodology, the speculative knowledge gained through the research can be considered valid in an expanded sense of that notion. By adding cultural studies to our more scientifically modeled research in technical writing, we can add discussions of power, politics, ethics, and cultural tensions to our understandings of what it is we do when we communicate. With an expanded idea of culture, we can expand our understanding of technical writing practice.

Works Cited

Aronowitz, Stanley. *Science as Power: Discourse and Ideology in Modern Society.* Minneapolis, U of Minnesota P, 1988.

Barton, Ben F., and Marthalee S. Barton. "Ideology and the Map: Toward a Postmodern Visual Design Practice." *Professional Communication: The Social Perspective.* Ed. Nancy Roundy Blyler and Charlotte Thralls. Newbury Park: SAGE, 1993. 49–78.

Bell, Arthur H., W. Tracy Dillon, and Harald Becker. "German Memo and Letter Style." *Journal of Business and Technical Communication* 9.2 (April 1995): 219–227.

Benjamin, Walter. *Illuminations.* New York: Schocken Books, 1968.

Boiarsky, Carolyn, et al. "Men's and Women's Oral Communication in Technical/Scientific Fields: Results of a Study." *Technical Communication* 42.3 (August 1995): 451–459.

Boisjoly, Roger, engineer with Morton Thiokol. *Company Loyalty and Whistle Blowing: Ethical Decisions and the Space Shuttle Disaster.* A talk at Massachusetts Institute of Technology. Videocassette. 7 January 1987.

Britt, Elizabeth, Bernadette Longo, and Kristin Woolever. "Extending the Boundaries of Rhetoric in Legal Writing Pedagogy." *Journal of Business and Technical Communication* 10.2 (April 1996): 213–238.

Burnett, Rebecca. "Conflict in Collaborative Decision-Making." *Professional Communication: The Social Perspective.* Ed. Nancy Roundy Blyler and Charlotte Thralls. Newbury Park: SAGE, 1993. 144–162.

Creighton, James Edward. "Special Introduction." *Advancement of Learning and Novum Organum.* Francis Bacon. New York: P. F. Collier & Son, 1900.

Dautermann, Jennie. "Negotiating Meaning in a Hospital Discourse Community." *Writing in the Workplace: New Research Perspectives.* Ed. Rachel Spilka. Carbondale: Southern Illinois UP, 1993. 98–110.

de Certeau, Michel. *The Practice of Everyday Life.* Trans. Steven Rendall. Berkeley: U of California P, 1984.

———. *Heterologies: Discourse on the Other*, translated by Brian Massumi. Minneapolis: U of Minnesota P, 1986.

———. *The Writing of History.* New York: Columbia UP, 1988.

Devitt, Amy J. "Intertextuality in Tax Accounting: Generic, Referential, and Functional." *Textual Dynamics of the Professions: Historical and Contemporary Studies of Writing in Professional Communities.* Ed. Charles Bazerman and James Paradis. Madison: U of Wisconsin P, 1991. 336–357.

Doheny-Farina, Stephen. "Research as Rhetoric: Confronting the Methodological and Ethical Problems of Research on Writing in Nonacademic Settings." *Writing in the Workplace: New Research Perspectives.* Ed. Rachel Spilka. Carbondale: Southern Illinois UP, 1993. 253–267.

———. *Rhetoric, Innovation, Technology: Case Studies of Technical Communication in Technology Transfers.* Cambridge: MIT P, 1992.

Dragga, Sam. "Women and the Profession of Technical Writing: Social and Economic Influences and Implications." *Journal of Business and Technical Communication* 7.3 (July 1993): 312–321.

Faigley, Lester. *Fragments of Rationality: Postmodernity and the Subject of Composition.* Pittsburgh: U of Pittsburgh P, 1992.

Foucault, Michel. *The Archaeology of Knowledge.* New York: Barnes & Noble, 1972.

———. *The Birth of the Clinic.* London: Routledge, 1973.

———. *Power/Knowledge: Selected Interviews and Other Writings 1972–77*, edited by Colin Gordon. Trans. Colin Gordon et al. New York: Pantheon, 1980.

Freed, Richard C. "Postmodern Practice: Perspectives and Prospects." *Professional Communication: The Social Perspective.* Ed. Nancy Roundy Blyler and Charlotte Thralls. Newbury Park: SAGE, 1993. 196–214.

Gordon, D. "Writing Culture, Writing Feminism: The Poetics and Politics of Experimental Ethnography." *Inscriptions* 3–4 (1988): 7–26.

Gramsci, Antonio. *Selections from the Prison Notebooks*, edited and translated by Quintin Hoare and Geoffrey Nowell Smith. New York: International Publishers, 1971.

Griffeth, Roger W., et al. "The Effects of Gender and Employee Classification Level on Communication-Related Outcomes: A Test of Structuralist and Socialization Hypotheses." *Journal of Business and Technical Communication* 8.3 (July 1994): 299–318.

Hall, Stuart, et al. *Culture, Media, Language: Working Papers in Cultural Studies, 1972–1979.* London: Hutchinson (Centre for Contemporary Cultural Studies), 1980.

Herndl, Carl G. "Teaching Discourse and Reproducing Culture: A Critique of Research and Pedagogy in Professional and Non-Academic Writing." *College Composition and Communication* 44 (1993b): 349–363.

Herndl, Carl, Barbara Fennell, and Carolyn Miller. "Understanding Failures in Organizational Discourse: The Accident at Three Mile Island and the Shuttle Challenger Disaster." *Textual Dynamics of the Professions: Historical and Contemporary Studies of Writing in Professional Communities.* Ed. Charles Bazerman and James Paradis. Madison: U of Wisconsin P, 1991. 79–305.

Herzberg, Bruce. "Michel Foucault's Rhetorical Theory." *Contending with Words: Composition and Rhetoric in a Postmodern Age.* Ed. Patricia Harkin and John Schilb. New York: Modern Language Association, 1991. 69–81.

Johnson, John Butler. "Two Kinds of Education for Engineers." *Engineering Education: Essays for English.* Ed. Ray Palmer Baker. New York: John Wiley & Sons, 1928. 67–86.

Katz, Steven. "The Ethic of Expediency: Classical Rhetoric, Technology, and the Holocaust." *College English* 54 (1992): 255–275.

Kleimann, Susan. "The Reciprocal Relationship of Workplace Culture and Review." *Writing in the Workplace: New Research Perspectives.* Ed. Rachel Spilka. Carbondale: Southern Illinois UP: 1993. 56–70.

Kossek, Ellen Ernst, and Susan C. Zonia. "The Effects of Race and Ethnicity on Perceptions of Human Resource Policies and Climate Regarding Diversity." *Journal of Business and Technical Communication* 8.3 (July 1994): 319–334.

Lather, Patti. "Fertile Obsession: Validity After Poststructuralism." *The Sociological Quarterly* 34.4 (1993): 673–693.

Lawson, Hilary. "Stories about Stories." *Dismantling Truth: Reality in the Post-Modern World.* Ed. Hilary Lawson and Lisa Appignanesi. London: Weidenfeld and Nicolson, 1989. xi–xxviii.

Lay, Mary M. "Feminist Theory and the Redefinition of Technical Communication." *Journal of Business and Technical Communication* 5 (1991): 348–370.

———. "Gender Studies: Implications for the Professional Communication Classroom." *Professional Communication: The Social Perspective.* Ed. Nancy Roundy Blyler and Charlotte Thralls. Newbury Park, CA: SAGE, 1993. 215–229.

———. "The Value of Gender Studies to Professional Communication Research." *Journal of Business and Technical Communication* 8.1 (January 1994): 58–90.

Leitch, Vincent B. *Cultural Criticism, Literary Theory, Poststructuralism.* New York: Columbia UP, 1992.

Lenzo, Kate. "Validity and Self-Reflexivity Meet Poststructuralism: Scientific Ethos and the Transgressive Self." *Educational Researcher* 24.4 (May 1995): 17–23.

Lyotard, Jean-François. *The Differend: Phrases in Dispute*, translated by Georges Van Den Abbeele. Minneapolis: U of Minnesota P, 1988.

———. *The Postmodern Condition: A Report on Knowledge.* Trans. Geoff Bennington and Brian Massumi. 1979. Minneapolis: U of Minnesota P, 1984.

McCarthy, Lucille Parkinson, and Joan Page Gerring. "Revising Psychiatry's Charter Document DSM-IV." *Written Communication* 11.2 (April 1994): 147–192.

Nelson, Cary, Paula A. Treichler, and Lawrence Grossberg. "Cultural Studies: An Introduction." *Cultural Studies.* Ed. Lawrence Grossberg, Cary Nelson, and Paula A. Treichler. New York: Routledge, 1992. 1–22.

Nicholson, Joel D., et al. "United States versus Mexican Perceptions of the Impact of the North American Free Trade Agreement." *Journal of Business and Technical Communication* 8.3 (July 1994): 344–352.

Paradis, James, David Dobrin, and Richard Miller. "Writing at Exxon ITD: Notes on the Writing Environment of an R&D Organization." *Writing in Nonacademic Settings.* Ed. Lee Odell and Dixie Goswami. New York: Guildford, 1985. 281–307.

Porter, James E. "The Role of Law, Policy, and Ethics in Corporate Composing: Toward a Practical Ethics for Professional Writing." *Professional Communication: The Social Perspective.* Ed. Nancy Roundy Blyler and Charlotte Thralls. Newbury Park: SAGE, 1993. 128–143.

Rorty, Richard. "Science as Solidarity." *Dismantling Truth: Reality in the Post-Modern World.* Ed. Hilary Lawson and Lisa Appignanesi. London: Weidenfeld and Nicolson, 1989. 6–22.

Sauer, Beverly A. "Communicating Risk in a Cross-Cultural Context: A Cross-Cultural Comparison of Rhetorical and Social Understandings in U.S. and British

Mine Safety Training Programs." *Journal of Business and Technical Communication* 10.3 (July 1996): 306–329.

Selzer, Jack. "Intertextuality and the Writing Process: An Overview." *Writing in the Workplace: New Research Perspectives.* Ed. Rachel Spilka. Carbondale: Southern Illinois UP: 1993. 171–180.

Smart, Graham. "Genre as Community Invention: A Central Banks' Response to Its Executives' Expectations as Readers." *Writing in the Workplace: New Research Perspectives.* Ed. Rachel Spilka. Carbondale: Southern Illinois UP: 1993. 124–140.

Stacks, Don W., and John E. Hocking. *Essentials of Communication Research.* New York: Harper Collins, 1992.

Stygall, Gail. "Texts in Oral Context: The 'Transmission' of Jury Instructions in an Indiana Trial." *Textual Dynamics of the Professions: Historical and Contemporary Studies of Writing in Professional Communities.* Ed. Charles Bazerman and James Paradis. Madison: U of Wisconsin P, 1991. 234–253.

Thralls, Charlotte. "Rites and Ceremonials: Corporate Video and the Construction of Social Realities in Modern Organizations." *Journal of Business and Technical Communication* 6 (1992): 381–402.

Tomlinson, Hugh. "After Truth: Post-Modernism and the Rhetoric of Science." *Dismantling Truth: Reality in the Post-Modern World.* Ed. Hilary Lawson and Lisa Appignanesi. London: Weidenfeld and Nicolson, 1989. 43–57.

Zak, Michele Wender. "'It's Like a Prison in There': Organizational Fragmentation in a Demographically Diversified Workplace." *Journal of Business and Technical Communication* 8.3 (July 1994): 281–298.

Zappen, James. "Francis Bacon." *Encyclopedia of Rhetoric and Composition: Communication from Ancient Times to the Information Age.* Ed. Theresa Enos. New York: Garland Publishing, Inc., 1996. 61–63.

Chapter Five

The Rhetorical Work of Institusions

Elizabeth C. Britt

> Any institution that is going to keep its shape needs to gain legitimacy by distinctive grounding in nature and in reason: then it affords to its members a set of analogies with which to explore the world and with which to justify the naturalness and reasonableness of the instituted rules, and it can keep its identifiable continuing form.
>
> —Mary Douglas, *How Institutions Think*

During the mid-1980s, the influence of the social perspective in technical communication moved our attention from the narrow confines of the text to the broader context of organizations. Researchers sought to understand the relationship between discourse forms and organizational structures and processes, launching a generation of empirical work on organizational communication (e.g., Dautermann; Doheny-Farina; Paradis, Dobrin, and Miller; Odell and Goswami; Pare). A decade later, the critical turn sparked an interest in complicating our ideas of the relationship between texts and contexts. Researchers began to attend to the workings of power and ideology, especially as a way to investigate the role of discourse forms in the creation, maintenance, and transformation of social relations. As part of this turn, scholars have called for increased scrutiny of institutions as something quite different from organizations. Bernadette Longo, for example, argues that the social perspective focuses too narrowly within organizational borders, ignoring the institutional forces that transcend them (12). These institutional forces legitimate some types of knowledge at the expense of other types, simultaneously constructing a particular reality and making this reality seem natural and inevitable. Institutions are coming to be seen as more than the setting within which rhetoric occurs; they are themselves "orchestrated" by rhetoric (Porter et al. 625). To avoid reifying their reality, Longo and others (e.g., Herndl) argue that we must connect the material practices of institutions to the discourse that supports them.

Institutions are becoming more prominent in analyses of discursive practices, especially in the work of scholars studying the rhetoric of medicine. For example, Judy Z. Segal's study of medical discourse connects textual features of journal articles to medicine's institutional character as paternal, atomistic, and interventionist. While Segal's stated goal is not to critique medicine, the interrogation of what she calls its "tenacity" (87) indicates an interest in how the institution maintains its dominance. McCarthy and Gerring's more explicitly critical approach examines the charter document of psychiatry, the *Diagnostic and Statistical Manual of Mental Disorders*, arguing that the manual reifies approaches to mental illness prominent in the institution of Western medicine. However, in both the calls for institutional analysis and in research responding to these calls, what is meant by institutions remains latent; the shift from "organizational" to "institutional" analysis is often treated as self-evident. Given that the concept of the institution is becoming more central to cultural studies projects in technical communication, some critical reflection on that concept and how we might use it is in order. What do we mean by the term *institution*? What do we gain by examining institutions? What have scholars in other fields said about institutions? What do we need to know about the particular institutions we are studying?

This chapter explores how researchers working at the intersection of cultural studies and technical communication might borrow from other fields—specifically, sociology and anthropology—to complicate our analyses of institutions. I suggest that we need to become familiar both with social and cultural theories of institutions and with the particularities of the specific institutions we study. To illustrate how such an approach might work, I analyze recent work on writing within insurance organizations, suggesting that a cultural studies approach to insurance as an important force in contemporary culture necessitates understanding not just how a particular insurance organization functions but also how insurance works as an institution. Although insurance is a ubiquitous and important feature of life in industrial and information economies, it is relatively invisible, even to scholars. As a mundane feature of everyday life, insurance provides a salient example of how institutions work in the background to both draw upon and shape what we perceive as common sense.

INSTITUTIONS AS TAKEN FOR GRANTED

Much technical communication research that examines discourse in context is concerned primarily with how texts and writers operate within organizations (see, e.g., Spilka). Institutions, if mentioned at all, are often seen as synonymous with organizations. Scholars adopting a critical approach appear to see institutions as different from organizations, although these differences are rarely foregrounded. This perspective is illustrated by a model offered by Porter et al. in their recent call for institutional critique in rhetoric and composition (622). The model, which represents potential sites for critique, is composed of three levels on a continuum from the most local to the most global. At the most local level are sites such as composition classrooms; at the most global level are *disciplines* (such as English studies) and *macroinstitutions* (such as the family or state). Situated in between are what Porter et al. call *microinstitutions/organizations* (such as literacy centers or writing labs). According to the authors, critiques aimed only at the local level are ineffective because they rely on "a kind of liberal, trickle-up theory of change" (617), while those aimed only at the global level fail because the entities they critique are monolithic and resistant to change. Porter et al. argue for critique aimed at the middle ground, the microinstitutions/organizations. This kind of critique extends beyond organizational borders—and beyond traditional organizational analysis—by attending to the power relations inherent in particular spatial and material conditions. The difference between traditional organizational analysis and institutional critique is signaled by the labeling strategy. Porter et al. label the top-level "macroinstitutions," using capitalization to signal that the entities belonging to this category are those ubiqituous abstract organizers of social life (e.g., Family, State, etc.). They then label entities belong to the middle level "microinstitutions/organizations," a strategy that calls attention to their power. By characterizing organizations as powerful entities—as a *kind* of institution—the authors open the door for critical analysis of them.

We can understand the impulse to see institutions as amenable to cultural critique by looking at how social theorists both differentiate between organizations and institutions and how they approach institutional and organizational analysis. Technical communication researchers are most concerned with what sociologists call "formal organizations." No consensus of opinion exists among social theorists on the meaning of these entities; they have been variously described as machines, organisms, brains, cultures, political systems, "psychic prisons," processes in a constant state of flux, and "instruments of domination" (Morgan). Social theorists disagree on how they function, both internally and with other social structures; the role of the individual in their functioning; the clarity and stability of their borders; and whether their meaning exists apart from human understanding. What seems common across most accounts, however, is a certain self-consciousness about their social relations (Silverman 8). That is, unlike the general patterns of social life (what sociologists call "social organization"), formal organizations are guided by explicit rules and clearly defined structures of authority (Silverman 8).[1] This self-consciousness is evident in Max Weber's classic definition of an organization as "a system of continuous activity pursuing a goal of a specified kind" (115) and is present even in more contemporary definitions that see organizations as cultures, for example, "as distinctive social units possessed of a set of common understandings for organizing action (e.g., what we're doing together in this particular group, appropriate ways of doing in and among members of the group)" (Louis 39).

Although the debate about institutions among social theorists is equally lively, a common assertion is that unlike formal organizations, which exhibit self-consciousness, institutions are marked by a certain taken-for-grantedness. Sociologists Peter L. Berger and Thomas Luckmann offer a model of how institutions emerge that explains how this taken-for-grantedness functions. Institutions emerge, they argue, with the inevitable habitual activity of social life (53). Over time, individuals develop patterns of interaction that become the predictable foundation or background of their social relations, the taken-for-granted arena within which they will create new ways of acting toward each other (57). When the patterns extend beyond the individuals who established them (to *types* of people performing *types* of actions), they have become institutionalized (54). The patterns now have a history that extends prior to the memory of the individuals taking part in them; they move from "the way I do things" to "the ways things have always been done" (54). This history rationalizes the patterns, which take on an objective reality that exists apart from the individual (59). As Berger and Luckmann write,

> An institutional world . . . has a history that antedates the individual's birth and is not accessible to his biographical recollection. It was there before he was born, and it will be there after his death. This history itself, as the tradition of the existing institutions, has the character of objectivity. The individual's biography is apprehended as an episode located within the objective history of the society. The institutions, as historical and objective facticities, confront the individual as undeniable facts. The institutions are *there*, external to him, persistent in their reality, whether he likes it or not. (60, emphasis in original)

With this distance, Berger and Luckmann argue, comes a need for the institution to justify itself. Without a biographical connection to the origins of the institution, individuals must be reminded of the rationale behind it. Although the institution does not seem less real, it can only continue to evoke this reality by socializing individuals through legitimating narratives (62). These narratives ultimately serve to reify the institution, to make it seem something other than a product of human creation (89). Legitimation occurs at the practical, everyday level and at the level of the cosmic worldview; a "symbolic universe" serves to bring together all the smaller pieces of the institution into a seemingly coherent whole (92). Legitimization is achieved, argues anthropologist Mary Douglas, by the "rhetorical resource" of the analogy, a device that hides the social origins of the institution and its justifying principles. For example, the analogy "female is to male as left is to right" grounds the political hierarchy of monarchy (e.g., people are to king as female is to male as left is to right) (49). These analogies connect institutionalized practices to nature, a move that grounds the social world in a realm understood as given and unquestionable. Each analogy serves as a "basic building block" that moves from the physical, observable world through the social structure (49). Rhetoric scholars have noted the role of discourse in achieving this invisibility for both science (Bazerman 14) and law (Fish).

Institutions, then, rely on rhetoric to justify and objectify their existence. They are also constituted by rhetoric. They perform their functions not only through material and technical mechanisms but also, as literary critic Vincent Leitch reminds us, through "ideological norms and protocols shaping the reception, comprehension, and application of discourse" (128). To imagine institutions in this way is to see them as active agents in the formation of culture, including the narratives that sustain ways of knowing and identities. It is also to see institutions as amenable to critique. As Porter et al. argue, "Though institutions are certainly powerful, they are not monoliths; they are rhetorically constructed human designs (whose power is reinforced by buildings, laws, traditions, and knowledge-making practices) and so are changeable. In other words, we made 'em, we can fix 'em" (611). Although enacting strategies for change is an essential part of institutional critique for Porter et al., that is not where I am headed here.[2] My goal is instead to illustrate how drawing on institutional theory in other fields can help us engage in criticism that moves beyond seeing the institution as a generic backdrop for discourse.

Insurance in Technical Communication Research

Organizational analysis in technical communication research can focus so much on the immediate contexts for discourse that it ignores the broader cultural forces that inform those contexts. Accounting for institutions means shifting attention to those structures that have become naturalized, or at least understood as necessary. Analyzing discourse as situated within institutions means interrogating how knowledge-making practices, ontological assumptions, and implicit values shape the reception or production of discourse. It also means examining how discourse functions (rhetorically and materially) within and between institutions.

To illustrate this approach, I focus on insurance, an institution with wide-reaching (though often invisible) influence. Only a small number of scholars—mostly sociologists, historians, anthropologists, economists, legal scholars, and political scientists—study insurance from a cultural studies perspective (see Baker and Simon). In the field of technical communication, the few studies involving insurance have focused solely on writers within insurance organizations and have not examined the relationship between these organizations and insurance as an institution. Studies of writers within insurance companies include a summary (but not an analysis) of the results of interviews with public relations writers (Pomerenke), recommendations about workplace writing based on a consulting experience (Timmons), an analysis of collaborative writing (Cross), and an evaluation of negative letters (Schryer). Two of these studies contextualize their analyses by examining the organizational context of the insurance company, paying attention to corporate structure, processes, and politics (Cross; Schryer). However, none of these studies connects the discourse of the organization to insurance as an institution. I turn to one of these studies, Catherine Schryer's "Walking a Fine Line: Writing Negative Letters in an Insurance Company," to illustrate how attending to institutional forces can open up opportunities for cultural critique. Schryer's primary focus is on improving the effectiveness of correspondence within this organization. While the article does not attend to insurance as an institution or engage in cultural critique, it does attend to issues of power and representation.

"Walking a Fine Line" is a report of research conducted as the result of a consulting experience, an exigence that helps to explain the main focus of the article. Schryer had been asked by managers at a large international insurance company to conduct writing workshops that would improve the effectiveness of letters informing clients that they had been denied long-term disability benefits. The managers were concerned with the "grammar and tone" of the letters as well as the fact that two-thirds of the decisions they conveyed were appealed. Before agreeing to conduct the workshops, Schryer convinced the company to allow her to examine the letters and the context of their production. She had two goals: one, to understand what constituted an "effective" letter, both by studying the letters themselves and by interviewing the writers and managers; and two, to understand the symbolic orders and forms of agency enacted by the letters. The article reports both her recommendations to the company for improving its negative correspondence and her critical assessment of negative correspondence in this context. Schryer argues that organizational processes and prefabricated templates compelled the writers to represent clients as passive and frozen in time. These representations were not in the best interests of either the clients or the writers, who had to face the clients' anger. Schryer's recommendations, emerging from this critique, aim at improving these letters by giving writers more resources and by helping them to represent readers as more capable of action.

Concerned with understanding negative letters in this particular context, Schryer creates an analytical framework that synthesizes genre and sociocultural theories. Drawing on rhetorical and linguistic approaches to genre as well as the work of Giddens, Bourdieu, and Bakhtin, she understands genres as emerging from the interaction between socialized individuals and specific organizations. Individuals are neither wholly free nor wholly constrained by genres, nor are genres static entities that individuals employ or reject wholesale. Rather, Schryer sees genres as flexible "constellations" from which individuals improvisationally select. These constellations are not neutral; as what Bakhtin calls a "chronotope," they instead reflect value-laden, taken-for-granted ideas about human agency relative to place and time. To investigate these constellations, Schryer borrows from Bourdieu's method of "social praxeology," a two-pronged approach that entails first examining the external resources that circumscribe individual thought and action and then investigating the understandings of individuals.

Schryer interprets Bourdieu's method of social praxeology narrowly. Loic J. D. Wacquant describes Bourdieu's method as a two-stage process that "weaves together a 'structuralist' and a 'constructivist' approach" (11). The first stage is an analysis of the "objective structures [that bear on] interactions and representations," while the second is an analysis of how the individuals involved perceive those structures (11). Defining which "objective structures" are relevant to a given project will determine the scope of the analysis. Schryer adopts a narrow focus by taking the texts themselves (rather than the organization or institution) as the objective structures. For Schryer, then, the first stage of social praxeology is to engage in "the close reading of texts that instantiate genres to describe and critique the strategies that some genres activate to represent power" (461). This close reading is achieved through readability analysis (which she believes will be persuasive to the managers), rhetorical analysis of the structure of the letters, and an analysis of the ideological assumptions embedded in the linguistic structures. Using readability analysis, Schryer learns that most of the letters are internally inconsistent in difficulty. The more difficult parts of the letters are sections that explain the clients' policies. Schryer engages in the second stage of social praxeology by placing this close reading in dialogue with material gleaned from her interviews. During the interviews, Schryer learns that the writers were required by the company to include these sections. Writers were allowed to make minor alterations to the policy sections, leading Schryer to call them both a resource and a constraint. She makes a similar claim about the templates used to compose the letters after seeing that all of the letters she examined followed the same structure. She learns from interviews that the writers and managers believed that negative letters must be structured so that the decision appears only after a buffer and an explanation. This belief was supported by the template that writers were required to follow. However, some letters had elaborated on this template by including specific details in the buffer (such as the name of the client's physician) and in the explanation section (such as analysis of particular pieces of evidence) to indicate that the client's case had been carefully considered. Both Schryer and the writers saw these letters as more effective than those that slavishly adhered to the model. Analysis of the linguistic features of all the letters revealed that more effective letters judged clients less harshly and represented them as more capable of action. After interviewing the writers and managers about how these decisions are made and conveyed, Schryer argues that this process encourages a chronotope in which clients are passive and frozen in time.

Through this dialectical method, Schryer situates the negative letter genre within its context. However, this context is limited to the immediate circumstances of the production of the letters, including the material resources available to the writers, legal requirements, financial constraints, and the influence of models. By substituting a close analysis of the letters and their immediate rhetorical context for a more thorough investigation of what Bourdieu calls *field*, Schryer misses an opportunity to investigate how insurance as an institution helps to create the conditions within which the negative letters function. Schryer calls Bourdieu's notion of field a way of understanding organizations or, "what we might formerly have called discourse communities" (456), a move that collapses disciplines, organizations, discourse communities, and social systems. The organization is part of the analysis only insofar as it bears on the decision making of the letter writers. This approach declines to interrogate, as Longo argues we must, how "technical writing participates in an economy of scientific knowledge and power within our culture" (17).

Within this narrowly defined context, the analysis of power relations is curtailed along two dimensions. First, the analysis is mostly limited to two sets of actors: the assessors/writers and the clients/readers. (A third group, the managers, figures into the article only briefly.) The analysis of the agency of the first group, which relies on the structuration theories of Giddens and Bourdieu, concludes that the writers experienced the models, templates, and requirements from other departments as both resources and constraints. In other words, the writers enacted and reproduced these genres, which in turn created frustration as they tried to create effective letters. This evidence, according to Schryer, "demonstrates the complex and contradictory operations of power within organizations" (472). The analysis of the agency of the readers draws on Bakhtin's theory that genres act as chronotopes (that is, that they enact a view of the world that places individuals in particular relations to time and space). Schryer writes: "Genres are forms of 'symbolic power' (Bourdieu 163) and could be forms of 'symbolic violence' (139) if they create time/spaces that work against their users' best interests" (460). To see whether readers are subjected to this violence, Schryer conducts a linguistic analysis of the letters, which reveals that clients are largely represented as immobilized by their health conditions and incapable of any activity other than speech acts, such as appealing the company's decision. The second way that the analysis of power relations is limited is that it does not connect the work of genres to the work of the institution. One of the indications that genres are enacting symbolic violence, according to Schryer, is when they are taken for granted as natural and normal. "One of the purposes of genre research," she writes, "should be to catch a glimpse of the 'chronotopic unconscious'" (460). While she investigates how this unconscious is manifested (by looking at the linguistic features of the letters), she does not connect that unconscious to the cultural work of insurance as an institution.

Interrogating Insurance as an Institution

Arguing that most of the rhetorical and linguistic approaches to genre do not engage in criticism, Schryer establishes critique as a goal of her article. Using social theory as a framework, she analyzes the power relations involved in the writers' decision-making, and she sheds light on the work of the negative letter in this context in representing clients to their disadvantage. This analysis grounds her recommendations for improving the conditions under which these writers compose negative letters and for representing the company's clients more positively. Schryer is able to engage in this critique partly because she distinguishes between "effectiveness" as the writers and managers see it and as she sees it. That is, she examines a letter not just for whether the client is likely to understand it (or appeal it) but also for its chronotopic representations of the client. While Schryer's focus on representation is driven by a critical bent, the end material goal is the same: both the writers/managers and Schryer have aimed at making clients more satisfied with decisions and thus less likely to appeal them. She is concerned with improving individual examples of the genre (a product focus) rather than on understanding how the genre as a whole functions within its institutional context (a process focus). This approach may make clients more satisfied, but it does not change or question the material relationship of clients to the institution. Investigating instead how the genre functions as part of larger institutional and cultural processes opens the door for critique.

Situating the negative letter genre in relation to these processes requires understanding insurance as an institution. Although the myriad kinds of insurance make it difficult (even for insurance scholars) to find commonalities among all of them, legal scholar Tom Baker has offered a broad definition of insurance as "a formal mechanism for sharing the costs of misfortune" (2). This definition focuses not on what insurance *is* but on what it *does*. This perspective, as explained by social theorist François Ewald, emphasizes insurance as a *technology*, as "an art of combining various elements of economic and social reality according to a set of specific rules" (197). Life insurance, for example, subjects data from mortality tables, physical examinations, and interview/confessional data about medical history, longevity, and lifestyle to the rules of statistical manipulation. These combinations are unique to particular insurance *forms* (such as marine insurance or social security); insurance is not the application of a single technology to any kind of loss (Ewald 197). What informs how these combinations arise, argues Ewald, are insurantial *imaginaries*—what rhetoricians might call constellations of topoi—that allow for the emergence of particular forms in a particular time and place. Ideas about what can be insured—and how—fluctuate with context; these determinations are based just as much on commonplace understandings as they are on actuarial tables and market interest (198). Given the complexity of insurance, it makes some sense to speak only of insurance institution*s* (in the plural). However, to the degree that technologies, forms, and imaginaries can cohere under a single term, we may also speak of *the* institution of insurance.

Institutions, according to Douglas, gain legitimacy through analogies to nature. Insurance gains legitimacy by its connection to risk, which has become naturalized, partly through its association with mathematical probability. As commonly understood, risk exists out there in the world; probabilities measure it, and insurance uses those probabilities to determine coverage and premiums as protection against it. Risk, according to Ewald, is fundamental to insurance, but its meaning differs from our commonsense understanding. Rather than applying to anything dangerous, a risk in insurance is "a specific mode of treatment of certain events capable of happening to a group of individuals" (199). A risk is not something that is dangerous or harmful in and of itself, but an event that has been subjected to various insurance technologies so that a form of insurance can be crafted to respond to it. In other words, a risk does not exist objectively in the world, but is created by the very technologies of insurance. As Ewald explains, "Nothing is a risk in itself; there is no risk in reality. But on the other hand, anything *can* be a risk; it all depends on how one analyzes the danger, considers the event" (199). Of course, Ewald is not arguing that dangers do not exist in the world, but that the identification of these dangers as *risks*—as subject to insurance—depends on the technologies and imaginaries currently at work. Following this logic means seeing insurance not as responding to risks but as producing them (199–200).

Thinking about insurance in this way opens it up to rhetorical and cultural analysis. Rather than being seen in naturalizing supply-and-demand terms, insurance can be understood as an argument about the nature of the world and the place of individuals in it. In the Foucaultian sense, it can be seen as a form of governance, "a form of activity aiming to shape, guide or affect the conduct of some person or persons" (Gordon 2). This governance is achieved in multiple intertwined dimensions, as detailed by Richard V. Ericson, Aaron Doyle, and Dean Barry in their extensive ethnography of insurance. Three of these dimensions deserve special attention for their potential to subject insurance to rhetorical and cultural criticism. First, as a "cultural framework for conceptions of time, destiny, providence, responsibility, economic utility, and justice" (6), insurance can be used as a lens through which to understand the commonplaces informing a particular cultural moment. By examining these commonplaces within the context of what Ewald calls the "insurantial imaginary" (198), we can use insurance to understand ourselves and the social world we have created. In other words, insurance can tell us something about *us* (e.g., what we deem as worthy of compensation and what we understand as being within our control). Two other dimensions open up the possibility for critique as well as analysis. Calling insurance "a social technology of justice" (6), Ericson et al. connect these commonplaces to the material world. Insurance not only reflects and shapes our thinking; it also has material consequences. For example, it provides monetary compensation to those who have experienced loss, it allows or forecloses certain kinds of livelihood and property ownership (for instance, whether one can run a business or buy a house), and it affects the shape of technology development (by allowing certain kinds of financial risks but not others). Finally, because of its function as a technology of justice, "insurance is therefore political, combining aspects of collective well-being and individual liberty in a state of perpetual tension" (6). To what extent are individuals responsible for their own misfortunes? What do members of society owe each other? How are these rights and obligations to be managed and by whom?

Within this framework, additional questions can be asked about the negative letter genre in Catherine Schryer's analysis. While her analysis does account for some activity within the insurance organization (e.g., the resources available to the writers and the necessity for legal language), it does not consider how the organization or the genre functions within insurance as an institution. Making this connection changes the analytical picture, making possible a range of critical responses.

One response is to examine the relationship between insurantial imaginaries and how clients are represented in the negative letters. Distinguishing between "effective" and "ineffective" letters, Schryer argues that the less effective letters represent readers as passive and frozen in time and that the more effective letters make the reader seem more capable of important actions: "The better letters simply had a higher level of activity assigned to all entities, including readers... [E]ffective letters often represented readers as capable of more than speech acts" (480), for example, of engaging in rehabilitation or some occupation. It is important to remember that all of these letters are negative letters; each is denying a claim for long-term disability. From the perspective of the managers, the difference between "effective" and "ineffective" letters seems to be whether the reader will make a costly appeal of the decision. The writers are also concerned about whether readers will make appeals, but at least part of their anxiety is about how the reader experiences the letter, regardless of whether an appeal is made. Both the managers and the writers also fear somewhat for the physical safety of the writers, who use pseudonyms when corresponding with clients. For Schryer, effectiveness has at least as much to do with the way the reader is represented; more effective letters are those that characterize readers more positively.

This concern with assessing effectiveness, while helpful for the instrumental purposes of the study, does not necessarily help us understand how the negative letter genre itself functions. A cultural critic might instead use the concern for effectiveness to formulate another set of questions: What is the significance of the relationship between the (presumed) likelihood of appeal and the way the reader is represented in negative letters? What does this genre tell us about the way clients understand their own responsibilities in relation to that of the insurer? These questions require first looking at the entire sampling of letters, which allows us to see tensions present *across* the genre. Schryer's linguistic analysis illustrates the range of activity levels and types that various entities (the policyholder, the insurance company, the assessor, medical conditions) are capable of. The policyholder (reader/client) is represented as more or less disabled, more or less capable of holding a job or getting better. The insurance company and the assessors (writers) are represented as more or less responsible for the decision to refuse benefits. The medical conditions are represented as more or less active, more or less amenable to rehabilitation. Schryer's analysis also illustrates a range of temporal positions for each entity, for example, whether the actions of the policyholder, insurance company, assessor, and so on are limited to the past or the future. This range of representations, understood by Schryer as rhetorical resources for writers, can also be understood as reflecting the tension in insurance between the individual and the collective. This tension is one of the most defining features of insurance. Ewald explains:

> Strictly speaking there is no such thing as an individual risk; otherwise insurance would be no more than a wager. Risk only becomes

> something calculable when it is spread over a population. The work of the insurer is, precisely, to constitute that population by selecting and dividing risks. Insurance can only cover groups; it works by socializing risks. It makes each person a part of the whole. Risk itself only exists as an entity, a certainty, in the whole, so that each person insured represents only a fraction of it. (203)

Yet the person does exist as part of the population, at least if he or she has been included in the risk pool. The individuals in a given risk pool can even be thought of as part of the insurance *product* itself, as the actions of each policyholder affect coverage and premiums (Ericson 5). Insurance can thus be seen as a "moral technology" by which individuals agree (through the purchasing of insurance) to take care of their responsibilities and (through the conduct of their lives) to call on the mutual assistance provided by the insurer only if previously agreed on circumstances prevail (Ewald 207; Ericson et al. 7). In other words, insurance demands that individuals, to a great extent, govern themselves through the processes and guidelines established by the insurer (Eriscon et al. 7). When we examine the negative letters in Schryer's study in this light, the agency attributed to various entities (policyholder, insurer, medical condition) takes on a different cast. The more "effective" letters, presumably those that managers and writers think are least likely to be appealed, are those that neither hold the policyholder fully responsible (through the language of blame) nor represent them as powerless. The workshops Schryer developed after conducting this research attempted to teach writers how to achieve this balance: "The style workshops centered on the theme of responsibility, that is, using actional constructions, especially transactives, to express what the company is expected to do and what the client is expected to do" (487). Even before the workshops, "[s]ome writers had located strategic ways to avoid blaming readers, and some had located ways in which readers could at least be represented as getting better, even if they were deprived of the long-term disability benefits they requested" (486). While the genre enforces individual responsibility (through the denial of benefits), it also represents the tension between individual and collective obligations on which insurance rests. Writers and policyholders alike seem to understand that policyholders must adhere to previously agreed on categories and procedures and that insurers must provide good reasons for their decisions. The role of each party in this relationship is taken for granted, as are the actuarial tables that define the categories by which decisions are made. More interestingly, the managers and writers believe that letters that attribute more agency to policyholders (but do not go so far as to blame them) will leave policyholders more satisfied. While we do not hear directly from policyholders, the implication here is that these representations reinforce notions of individual responsibility, increasing their willingness to shoulder these burdens alone.

The tension between individual and collective responsibility is also indicated by the writers' characterization of the "fine line" they walk in composing these letters. According to one of the writers, this tension is created because "the person in me wants to be sympathetic, but the assessor in me has to make a business decision" (Schryer 446). The "fine line" is understood as being between emotion and business, personal feelings and economic realities. However, it is important to distinguish the business of insurance from other types of business. As a moral technology, for example, insurance both governs and demands self-governance. With self-governance comes the potential for fraud, for individuals either to make false claims or to claim benefits greater than their situation requires. Because fraud is built into this technology, insurance has built-in checks and balances to discourage it.[3] One of these checks is the claims process. As Ericson et al. explain, assessors are embedded in a "micro-negotiation of political economy and fiscal responsibility" in which fraud is part of the institutional fabric (344): "The assumption is that first, the claimant presents an inflated claim knowing that the adjuster will try to minimize it. Second, the claimant knows that if she persists to a reasonable degree, she may be offered a nuisance payment because the adjuster needs to close the file efficiently. Third, if she is reasonable in finessing her claim, the claimant may be able to garner extra compensation for her reasonableness" (344). While Schryer's writers are also the claims adjusters—that is, they make the decisions that are being conveyed in the letters—only brief attention is paid to this part of their role. We do learn that the writers analyze the claims of policyholders to determine if their cases warrant coverage under the definitions of a particular plan. We also learn that the writers, most of whom do not have college degrees or any medical training, use a template that requires them to distill the (usually voluminous) medical files into the medically significant categories of signs, symptoms, and categories. While Schryer notes (but does not fully interrogate) this categorization as a knowledge-making process, neither the role of insurance as an institution nor the organization as a business is considered. Understanding the role of fraud—and its entanglement with notions of individual and collective responsibility—in insurance helps to frame the writers' dilemma in different terms. The "fine line" that the writers walk is constructed not only by the dichotomy between the writers' personal feelings and their professional duties, but also by the nature of the job—to minimize claims paid—within the political economy of insurance. This political economy is driven not just by the desire to maintain profits (as with any business) but by the role of insurance to both collectivize and individualize responsibility.

The role that any policyholder expects to play within the political economy of insurance is influenced by the insurantial imaginary at work. While insurance at any cultural moment functions as a kind of justice (Ewald 206), the shape of that justice can differ. For example, Baker and Simon have noticed a shift at the end of the twentieth century in the balance between the individual and the collective ("Embracing Risk"). Within the developed world, most of the twentieth century was characterized by expanding social insurance systems (3). The decades since the 1980s, however, have been marked by systems and forms that encourage individual risk-taking as a way of reducing burdens on the group (3). The most prominent example of this shift is the debate over privatizing Social Security, begun in earnest during the 2000 presidential campaign when both major candidates favored a move away from the collective model that had been in place for decades (4). The trend toward embracing risk does not stop with formal insurance systems, however. The increased numbers of people playing the stock market (especially as day traders), gambling, and participating in extreme sports (5) indicates that the imaginaries informing our insurance systems also undergird other cultural formations. Using this shift as a lens, the objects of Schryer's study take on increased cultural significance. The writers and policyholders are not only embedded in interpersonal and organizational relations; they are also located within a cultural system whose foundational assumptions are in flux.

The Institution as Rhetoric

Organizational approaches to technical communication—especially those that envision organizations as cultures—have given us insight into the production and use of texts. When these approaches consider power relations and ideology, they can enter into the realm of cultural critique. To avoid treating organizations as isolated entities apart from other organizations and broader forces (economic, political, and cultural), researchers also need to interrogate institutions. Such analyses can help illuminate texts and the actions of individuals as part of broader structural formations in addition to localized structures and processes. Yet institutions do not exist in isolation either. Because they "tend to 'hang together'" (Berger and Luckmann 63), institutions can—and should be—examined as cultural agents entangled with other institutions. For example, while I focused on the institution of insurance in my analysis of Schryer's study, insurance is embedded in a network of institutions, law and medicine among them. The connections among these institutions should be subjected to scrutiny.

This chapter has also attempted to illustrate that while institutions can be seen as contexts for discourse, they can also be seen as discursive formations, as rhetorics that can be subjected to criticism. This way of understanding institutions allows us to treat them as "unit(s) of analysis" (Porter et al. 632) rather than as passive containers within which discourse occurs. For example, institutions can be analyzed, broken down into their constituent parts, by translating the ideas of social theorists into the language of rhetoric. (In this article, I gained analytic purchase by translating Ewald's insurantial imaginaries as constellations of topoi.) Approached as objects of rhetorical inquiry, institutions can be examined as agents in cultural narratives, as entities that both rely on and influence what we perceive as reality.

Taking such an approach means seeing that technical communication does not occur just at the intersection of product and consumer or across an organization. Technical communication is the means by which institutions define themselves and conduct their cultural work. To investigate this work, we must become more familiar with social and cultural theories of institutions, such as those cited in this chapter. Equally important is that we commit ourselves to understanding particular institutions by connecting with scholars in other fields who study them. Because of the daunting nature of this task, a researcher may need to specialize in a single institution, even though such specialization may preclude in-depth analyses of the interrelations among institutions. With this background, we can begin our critical interrogation: What does this institution do (culturally, materially, symbolically)? What forms of knowledge and modes of discourse does this institution recognize as legitimate and illegitimate? Whose interests are served and whose ignored? How does this institution justify its legitimacy? How does rhetoric help this institution do its work?

NOTES

1. It is important to note that Silverman problematizes the distinction between formal organizations and the rest of social life, arguing that neither can be understood in isolation and that their borders are not rigid (23).

2. Porter et al. argue that "[t]o qualify as institutional critique, a research project has to actually enact the practice(s) it hopes for by demonstrating how the process of producing the publication or engaging in the research enacted some form of institutional change" (628). Such an interventionist stance may make sense when the changes needed are clear and when the researcher is in a position to enact change. But I disagree with the assertion that work that does not enact change cannot count as critique. Work that analyzes and explains cultural processes (but stops short of intervention) need not take an accommodating stance. In fact, careful critical analysis is a necessary prerequisite to any meaningful change.

3. For an introduction to "moral hazard" (the idea that insurance encourages reckless behavior), see Tom Baker (1996), "On the Genealogy of Moral Hazard," *Texas Law Review* 75: 237–292.

Works Cited

Baker, Tom. "Insurance and the Law." *Encyclopedia of the Social and Behavioral Sciences* 11 (2002): 75–87.

———. "On the Geneology of Moral Hazard." *Texas Law Review* 75 (1996): 237–292.

Baker, Tom, and Jonathan Simon. "Embracing Risk." *Embracing Risk: The Changing Culture of Insurance and Responsibility.* Ed. Tom Baker and Jonathan Simon. Chicago: U of Chicago P, 2002. 1–25.

———. *Embracing Risk: The Changing Culture of Insurance and Responsibility.* Chicago: U of Chicago P, 2002.

Bazerman, Charles. *Shaping Written Knowledge: The Genre and Activity of the Experimental Article in Science.* Madison: U of Wisconsin P, 1988.

Berger, Peter L., and Thomas Luckmann, *The Social Construction of Reality: A Treatise in the Sociology of Knowledge.* New York: Anchor, 1967.

Bourdieu, Pierre, and Loic J. D. Wacquant. *An Invitation to Reflexive Sociology.* Chicago: U of Chicago P, 1992.

Burchell, Graham, Colin Gordon, and Peter Miller, eds. *The Foucault Effect: Studies in Governmentality.* Chicago: U of Chicago P, 1991.

Cross, Geoffrey A. "The Interrelation of Genre, Context, and Process in the Collaborative Writing of Two Corporate Documents." Spilka 141–157.

Dautermann, Jennie. "Negotiating Meaning in a Hospital Discourse Community." Spilka 98–110.

Doheny-Farina, Stephen. "Writing in an Emerging Organization: An Ethnographic Study." *Written Communication* 3 (1986): 158–185.

Douglas, Mary. *How Institutions Think.* Syracuse: Syracuse UP, 1986.

Ericson, Richard V., Aaron Doyle, and Dean Barry. *Insurance as Governance.* Toronto: U of Toronto P, 2003.

Ewald, François. "Insurance and Risk." Burchell, Gordon, and Miller 197–210.

Fish, Stanley. "The Law Wishes to Have a Formal Existence." *The Fate of Law.* Ed. Austin Sarat and Thomas R. Kearns. Ann Arbor: U of Michigan P, 1991.

Gordon, Colin. "Governmental Rationality: An Introduction." Burchell, Gordon, and Miller 1–51.

Herndl, Carl. "Teaching Discourse and Reproducing Culture: A Critique of Research and Pedagogy in Professional and Non-Academic Writing." *College Composition and Communication* 44 (1993): 349–363.

Leitch, Vincent B. *Cultural Criticism, Literary Theory, Poststructuralism.* New York: Columbia UP, 1992.

Longo, Bernadette. *Spurious Coin: A History of Science, Management, and Technical Writing*. Albany: SUNY P, 2000.

Louis, Meryl Reis. "Organizations as Culture-Bearing Milieux." *Organizational Symbolism*. Ed. Louis Pondy, Peter Frost, Gareth Morgan, and Tom Dandridge. Greenwich, CT: JAI, 1983. 39–54.

Morgan, Gareth. *Images of Organization*. 2nd ed. Thousand Oaks, CA: SAGE, 1997.

Odell, Lee, and Dixie Goswami, eds. *Writing in Nonacademic Settings*. New York: Guilford Press, 1985.

Paradis, James, David Dobrin, and Richard Miller. "Writing at Exxon ITD: Notes on the Writing Environment of an R&D Organization." *Writing in Nonacademic Settings*. Ed. Lee Odell and Dixie Goswami. New York: Guilford, 1985. 281–307.

Pare, Anthony. "Discourse Regulations and the Production of Knowledge." Spilka 111–123.

Pomerenke, Paula J. "Writers at Work: Seventeen Writers at a Major Insurance Corporation." *Journal of Business and Technical Communication* 6 (1992): 172–186.

Porter, James E., Patricia Sullivan, Stuart Blythe, Jeffrey T. Grabill, and Libby Miles. "Institutional Critique: A Rhetorical Methodology for Change." *College Composition and Communication* 51 (2000): 610–642.

Schryer, Catherine F. "Walking a Fine Line: Writing Negative Letters in an Insurance Company." *Journal of Business and Technical Communication* 14 (2000): 445–497.

Segal, Judy Z. "Writing and Medicine: Text and Context." Spilka 84–97.

Silverman, David. *The Theory of Organisations*. London: Heinemann, 1970.

Spilka, Rachel, ed. *Writing in the Workplace: New Research Perspectives*. Carbondale: Southern Illinois UP, 1993.

Timmons, Theresa Cullen. "Consulting in an Insurance Company: What We as Academics Can Learn." *The Technical Writing Teacher* 15 (1988): 105–110.

Wacquant, Loic J.D. "Toward a Social Praxeology: The Structure and Logic of Bourdieu's Sociology." *An Invitation to Reflexive Sociology*. Pierre Bourdieu and Loic J. D. Wacquant. Chicago: U of Chicago P, 1992.

Weber, Max. *Basic Concepts in Sociology*. 1962. New York: Citadel, 1993.

ACKNOWLEDGMENTS

Thanks to Barry Cohen for pointing me toward literature on the sociology of institutions, to Terese Guinsatao Monberg and Elizabeth Shea for helping me conceptualize my argument, and to Bernadette Longo, Blake Scott, and Katherine Wills for helpful suggestions on drafts.

Chapter Six

The Study of Writing in the Social Factory

Methodology and Rhetorical Agency[1]

Jeffrey T. Grabill

In this chapter, I take as my task a consideration of how research methodology in technical and professional writing might understand its relationship to cultural studies. One of the general arguments of this book is that technical and professional writing research needs to be more openly critical of unjust conditions, thereby enabling both deeper understandings of the relationships between cultural and technical practices and the ability to help intervene when possible. However, there is an assumption that professional and technical writing *needs* cultural studies, that cultural studies provides something that is currently missing and cannot be found either "inside" the field or elsewhere (e.g., rhetorical theory). I'm not sure this is true. My general view is that the interplay between studies of technical and scientific discourse in specific contexts and rhetorical studies of culture is necessary, but that this interplay maps a research terrain that redirects both cultural studies and technical and professional writing.

As I hope to make clear, both cultural studies theorists and critical researchers in technical and professional writing share the "problem space" of how to understand and create possibilities for change. I think research can help create these possibilities. However, I will argue for a specific way of understanding research methodology. I will use that methodological framework to foreground research practices as the mechanism for rhetorical agency within a research project. I have great hope for these largely unarticulated research practices. To be effective, however, they must be informed by the rhetorical study of culture (which is located at the boundary between rhetoric and anthropology), and a rhetorical understanding of institutional practices (which can be found in technical and professional writing). This is a great deal to ask of any research methodology, but it seems to me that researchers in technical and professional writing are well-positioned to address deep problems of rhetorical agency, and to address them in precise ways within a range of situated institutional contexts.

My approach will be to begin by representing the critical research discourse in technical and professional writing and a methodological framework that I find useful for foregrounding research practices. I will then move to a consideration of cultural studies in relation to rhetorical studies of culture. Building on these discussions, I consider how research practices can become transformative rhetorical acts. Throughout, I will consider questions of research location and audience as ways of pushing methodology toward critical possibility.

METHODOLOGY AND CRITICAL RESEARCH IN TECHNICAL AND PROFESSIONAL WRITING

Researchers in technical and professional writing have been engaged in a healthy discussion of research methodology for some time. Blyler as well as Herndl and Nahrwold provide the most comprehensive reviews of what they call descriptive and instrumentalist research in technical and professional writing. Their critique is that much of the research conducted in technical and professional writing merely describes the workplace writing scene. Too little engages in critique and intervention.[2] The problem with this situation is that descriptive research also typically supports, either explicitly or implicitly, the interests of those who sponsor research—either a funding agency or the institution that grants permission or both. While there isn't necessarily anything wrong with this situation, critics of descriptive (and instrumental) research argue that it often leaves in place and unquestioned—indeed often unexamined—ideological and cultural practices that are important for more complex understandings of discursive practices and the institutions that warrant them.

Cynthia and Richard Selfe note an important tension in conceptualizations of technical communication between private (e.g., the corporation) and public ownership, location, and use (330). For them, this tension raises a question of responsibility, namely whose interests do we serve through research and teaching in technical communication? Critical researchers in technical and professional writing have answered this question in various ways. Carl Herndl, for example, has examined questions of power and has been most consistently aware of how postmodern theory, critical theory, and rhetoric have impacted research and writing in anthropology and sociology, and in turn, how changes in those disciplines have impacted research on writing (see "Writing Ethnography," "Teaching Discourse," and "The Transformation"; see also Doheny-Farina). For Blyler, the critical perspective is "concerned not with describing and explaining a given aspect of reality, but rather with discovering what that aspect of reality means to social actors..." (36). It means, as well, that the goals of research should be emancipation, empowerment, social change. Thus the purposes and goals of research have changed (for some) in an effort to acknowledge that which was always true: we "take sides" in our research. So how do we do this in ways which are ethically and politically sound (or at least justifiable)? While not everyone agrees with the concerns expressed by Herndl, Blyler, and others (see, for example, Charney; Cooper; Barton), there is little question that research in technical and professional writing has taken a critical turn. At the very least, sound arguments have been made that cannot easily be dismissed. The critical turn in professional writing research plays out the implications of long-standing methodological issues: the neutrality and objectivity of the researcher; the goals and purposes of research; and the rhetorical nature of research itself. The result is an openly political research stance that incorporates into study design important issues of power and position.

These examinations of methodology are useful; however, they tend to be *about* methodology, not discussions of methodological practice. I am arguing for attention to practice. And so, to accomplish my goals in this chapter, let me move to Patricia Sullivan and James Porter's critical research framework (see fig. 6.1). Perhaps the most significant move they make in this framework is to distinguish between "method" and "methodology," with methodology serving as a much larger term encompassing ideology, practice, and method.

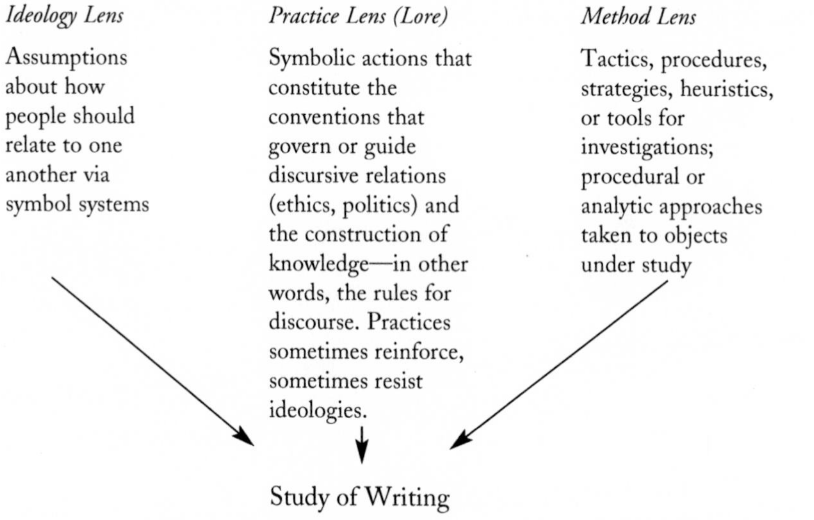
Ideology Lens
Assumptions about how people should relate to one another via symbol systems
Practice Lens (Lore)
Symbolic actions that constitute the conventions that govern or guide discursive relations (ethics, politics) and the construction of knowledge—in other words, the rules for discourse. Practices sometimes reinforce, sometimes resist ideologies.
Method Lens
Tactics, procedures, strategies, heuristics, or tools for investigations; procedural or analytic approaches taken to objects under study
Study of Writing

FIGURE 6.1. Sullivan and Porter framework for research methodology. Reprinted from *Opening Spaces: Writing Technologies and Critical Research Practices* by Patricia Sullivan and James E. Porter. Copyright © 1997 by Greenwood Publishing Group, Inc. Reproduced with permission of Greenwood Publishing Group, Inc., Westport, CT.

The distinction between "method" and "methodology" is important. In Sullivan and Porter's model, something like "ethnography" is not a method. It is a methodology, which means, more precisely, that a particular approach or theory of ethnography has an ideological component (a theory of human relations), a practice component (a notion of how people actually constitute relations with each other), and a method component (tools and so forth). Such a model allows us to discuss various ethnographic methodologies and to account for their differences. The importance of this way of thinking about research is that it enables a way of talking about differences within traditions, and it prevents the separation of methods from ideology—the separation of modes of inquiry from arguments about values that, for some, confuse and slant the research scene and for others make research worth doing.

The space of methodology is difficult, contested, and uncomfortable. Herndl and Nahrwold describe it as a "disciplinary situation in which both a modernist faith in method and a postmodern suspicion of grand narratives . . . exist within the same institutional space" (260). Herndl and Nahrwold argue that a researcher's personal and political commitments and purposes drive epistemological and methodological choices, which, they note, is different from how methodological choices are typically made (or at least different from how researchers typically write about the choices they make). Often epistemological and methodological choices shape the research scene. Sullivan and Porter's framework allows us to think somewhat differently. It allows us to see methodology as simultaneously shaped by a number of issues. Disciplinary histories, values, and training impact the researcher via the ideologies and the methods that shaped one's training. Personal commitments and positions are fully penetrated by ideologies and also constitute the ground for the research practices available and possible.[3] The combinations and vectors of influence are many. The point is to understand that methodology must attempt to account for them. We begin from the premise, therefore, that research methodology in technical and professional writing operationalizes—throughout the research process—an articulation of ideologies, practices, and methods.

My goals in this opening section are modest: to gloss the conversation about critical research within technical and professional writing and to lay out a way of thinking about research methodology that is expansive in its scope yet precise in terms of informing specific study designs. I will come back to this framework later. I want to move to a consideration of cultural studies and the concept of culture.

CULTURAL STUDIES AND RHETORICS OF CULTURE

Let me begin with Stuart Hall's admonition that cultural studies is not and never has been "one thing" ("Emergence" 11). Hall's statement is confirmed by the work in this volume, which draws from a broad range of disciplines and traditions. Certainly "cultural studies" is diverse, and because of this diversity, it is a contested term. Thus Nelson, Treichler, and Grossberg argue that it matters how cultural studies is defined and that it is a field, despite the fact that the approaches scholars take are indeed diverse (3–4). In my discussion here, I work within the field's most obvious definition and history, the trajectory that begins with the Centre for Contemporary Cultural Studies in Birmingham and with its rich tradition of intellectual and activist work. Nearly twenty-five years ago, Stuart Hall traced two paradigms that emerged from this trajectory, one "culturalist" and another "structuralist" ("Cultural Studies"). Hall views the culturalist paradigm as dominant and as methodologically more particular and situated in its approach to the study of culture. The work Hall cites within this paradigm is historiographic and anthropological. The structuralist paradigm is more abstract and explicitly concerned with ideology and discourse. Both work at the level of "forms and structures," and indeed, I would argue that these forms and structures are essential to programs for social change within this tradition. Over the last twenty-five years, it seems that what Hall calls a "structuralist" approach (not to be confused with structuralism per se) has become the dominant paradigm, certainly within English studies. As cultural studies has been filtered through English, its methodological legacy has come to meet certain ways of reading cultural texts.

If we think of cultural studies as offering "a bridge between theory and material culture" in a way that seeks to transform both (Nelson, Treichler, and Grossberg 6), and if we think of cultural studies theory as encompassing both production and consumption of cultural artifacts (which Nelson, Treichler, and Grossberg assert), then the intellectual and activist space cultural studies creates for itself is large. Indeed, serious attention to cultural conditions and to both production and reception distinguish cultural studies from many theoretical enterprises in the humanities. However, the claims made for its *methodological* reach are problematic. As I see it practiced within English studies, cultural studies is a way of reading. The question I consider here is what researchers in technical and professional writing can take from this form of cultural studies to transform their work, particularly since technical and professional writing has developed its own discourse on critical research practices outside of this tradition. While discourse analysis has long been a form of research in technical and professional writing, I'm suspicious of the claim that another way of reading technical and scientific texts—even a radically different approach—offers a meaningful alternative. Moreover, I doubt very much that these ways of reading, given the locations (academic publications) and audiences (other academics) for the research, constitute much of a transformative practice. My concern, in other words, is that alternative ways of reading texts cannot fill the activist space of cultural studies, and that, more important, the production of new readings for academic audiences is only marginally useful.[4] I am also asserting that empirical methodologies are potentially transformative in ways that textual interpretation cannot be. My concern is not so much with texts as an object of study as it is with the narrow range of texts studied, the methodologies utilized, and the limited circulation of the products of interpretation.

However, it is equally true that the concept of "culture" as an analytical focus is largely absent from research in technical and professional writing, and it is also true that while there is an established critical discourse within technical and professional writing, issues of culture, class, and difference are not well developed.[5] There is a real need to understand the cultural conditions that help shape the production and consumption of technical and professional writing. There is an important connection to be made, then, at the point at which the rhetorical study of cultures and institutions meet, a space neither cultural studies scholars nor technical and professional writing researchers visit much (for an exception, see Longo). It is within this space that one finds notions of culture from anthropology that are germane for technical and professional writing, and an analytical space that is dependent (methodologically at least) on sophisticated studies of culture, rhetorical theory, and the core concerns of technical and professional writing.

Ralph Cintron writes that "the process of making; made things as cultural displays or performances; the economic, social, and political contexts of made things; the circulation of such things through the imaginations of a community and a culture" drive his work in *Angels' Town*, an ethnography of acts of making in a Latino/a community. It is common in contemporary anthropology to approach "culture" in similar ways. Culture is not found, wholly formed, it is created in at least two ways: by the participants in/of the culture and by the researcher making sense of the cultural moment. Discussing the work of Paul Willis, whose studies of class and culture in England have become touchstones in cultural studies, Foley notes that for Willis, everyday groups invent cultural practices. He notes as well that this sense of emergent cultural practices contrasts with anthropological notions of a "historical cultural tradition that is passively inherited" (139). Foley, who is reaching for new ways to understand culture and class as both dynamic and material, likes the way in which Willis injects class into notions of cultural distinctiveness, thereby linking both present struggles and ongoing processes of cultural construction with historical traditions.

Outlining an even deeper sense of cultural "making," Kathleen Stewart characterizes culture (and the research acts of sense making) as rather incoherent, something that is physical and material, rhetorical and performative, psychological and emotional. Something that is indeed a representation, one that clearly extends beyond the way an anthropologist inevitably "fixes" her sense of culture. For Stewart, the "makers" of a culture—again both participants and researcher—must inhabit both physical and narrative spaces. Thus, she writes, "Picture 'culture' in the coal camps [one site of her work in Appalachia], then, not as a finished text to be read or as a transparent 'object' that can be abstracted into a fixed representation but as a texted interpretive space in itself—a space produced in the slippage, or gap, between sign and referent, event and meaning, and gathered into performed forms and tactile reminders" (26–27). Julie Lindquist builds on this notion of culture as narrative. Early in her book on the rhetorics of class in a working-class bar, she writes that because ethnography "approaches social interaction heuristically, any interpreted account of behavior presupposes a theory of culture" (4). And so if that theory of culture is a theory of narrativity, culture "must be understood as relational as well as distinctive, as a site of action and reaction" (5). Echoing both Foley and Stewart, Lindquist goes on to write that "Implicit in my claim to take as my subject 'working-class culture' is the assumption that shared cultural experiences (and the narrative processes and products of these experiences) are linked to material conditions, that what happens at the local level manifests what is structural and systemic" (5). In terms of what culture is and how it means, my gloss of these anthropologists and ethnographers frames some key arguments about the nature of "culture," such as the notion that a culture is a function of an ongoing process of construction, that a researcher seeking to understand that culture must "fix" it momentarily, and that the process of writing ethnography is deeply problematic and rhetorical.

Indeed, given the preoccupation with the rhetorics of culture—that is, with how we come to know and create "culture"—it isn't surprising that Foley's discussion of Paul Willis connects Marx and Habermas in an attempt to articulate a material and discursive method for understanding the creation of culture. Foley focuses on Marx's notion of "sensuous labor," or "what most anthropologists would call shared, expressive, 'nonproductive' cultural practices such as eating, dancing, singing, and joke making" (146). Foley goes on to claim that Habermas further develops Marx's "idea of alienated labor activity into a general theory of communicative action" (147). This strikes me as a remarkable move to imagine Habermasian communicative action as (alienated) labor. Regardless of what one thinks of this leap, Foley creates interesting space at the intersection of cultural studies and technical and professional writing. Thus Foley writes: "The question then becomes how to study 'labor' defined in the broader sense as communication/miscommunication. Such studies would focus on 'communicative labor' in the 'social factory' of everyday life and in 'nonproductive' cultural institutions such as schools, media, and families" (147). Foley is calling for Habermasian studies of communicative labor that would explore how processes of communication and miscommunication (Habermasian) produce class positions and resistances to those class positions. Foley has put his finger on something terribly important. Setting aside the preoccupation with truth claims implied by the communication/miscommunication dynamic, we might still understand technical and professional writing as a type of communicative labor, acts of making that take place in diverse cultural and institutional contexts, that circulate widely, and that do much more than enable functional tasks. To approach such understandings, we need rhetorical studies of culture, not textual studies of (ideological, cultural, rhetorical) production.

So what might these studies look like? Well, first we need to look for models of good cultural *and* rhetorical theory. Another way that Cintron describes his work is as a "project in the rhetorics of public culture or the rhetorics of everyday life" (*Angels' Town* x). Such a project concerns itself with makers and various acts of making, performative human actions such as clothing, hairstyles, speech, writing, and gesture. Rhetoric as a way of framing these acts of making is important because rhetoric is "interested in structured contentiousness that organizes, albeit fleetingly, a community or a culture" (x). Methodologically, Cintron's work, as he describes it, exists at an intersection between sociocultural anthropology and rhetoric (for more on this connection, see "Gates Locked"):[6]

> I see a project in the rhetorics of public culture as a somewhat new approach, one that adopts the fieldwork methods traditional to sociocultural anthropology and blends these with the cultural critique now common among critical ethnographers and theorists, and picks up as well ideas from an entire lineage of rhetorical theorists stretching to classical Greece and Rome. (xi)

Similarly, Julie Lindquist has recourse to rhetoric because of its analytical power for working in the "interpretive domain" between "class and culture, between the structural and the phenomenological" (8). Rhetoric not only allows her to understand how patrons at the Smokehouse Inn are persuasive, it is for her an approach to understanding class and culture itself. Rhetoric, she writes, "helps us to account for how the Smokehouse conceives of itself as a political culture, implying strategic positioning, public presentation, and persuasion" (8). Rhetoric is a heuristic for understanding the "microprocesses of social change," which in Lindquist's work are fundamentally communicative processes.

There is much that technical and professional writing might learn by embracing a rhetoric of culture in how we design research to understand the contexts of our work, in particular the ideological commitment to locating resistances and the procedural commitment to understanding culture as performative. The burden this places on technical and professional writing research is to account for cultural production as always already taking place within the diverse contexts where technical and scientific discourse is produced and where this discourse circulates.

Rhetorics of Institutions and the Space of Agency

Much technical and professional writing research has proceeded utilizing thin notions of culture—the "organization" serves as a static backdrop or unexamined context for many studies. The invention of that organizational culture is not often the subject of research; it is not a rhetorical possibility given the methodological frames employed. However, many scholars in cultural studies and critical researchers in technical and professional writing struggle with the same problem: how to locate and understand agency. Granted, this problem is articulated differently; however, it is *located* similarly. That is, contemporary institutions are the scene of struggle. For some, this scene is portrayed as postmodern or postindustrial capitalism. For others, the scene is the corporation. For both, the problem of agency is the problem of acting within systems of decision-making marked by organizational, epistemological, and discursive complexity. These institutional systems thoroughly penetrate our lives (Porter, Sullivan, Blythe, Grabill, and Miles). There are methodological possibilities in such a situation that enable new understandings of rhetorical agency.

I want to begin with the very problem of location. If we think of location spatially, then both where we conduct our research and with (and for) whom we work become foregrounded. Foley's call for research that examines communicative labor in public life and social institutions is a call I am willing to echo. But who will do this work? And how can it be done in ways that are intellectually sound and that enable agency? These are interesting questions that I think technical and professional writing researchers are well-positioned to help answer if we locate our research differently, both literally (studying a wider range of institutions) and intellectually (taking culture as a central problematic). Location also demands a methodological framework as robust as Sullivan and Porter's because it pushes on the relevance of more common ways of thinking about research (as a function of ideology or method or as a function of questions and disciplinary concerns) and underlines the importance of methodological terms that are commonly ignored, such as practice.

My use of the term "practices" differs slightly from the use suggested by Sullivan and Porter (see fig. 6.1). For Sullivan and Porter, practices are conventional, the rules that govern discourse. But they also have the suggestion of activity. I would like to preserve their notion that practices are conventions for how relations are constructed through research, but it seems important to acknowledge that these conventions are operationalized as actions, performances. Therefore, the sense of practice that I articulate in this chapter is more like the tactical practices of de Certeau—a *metis*-like knowing how, or situated maneuvers that are knowledge-rich. In this sense, my use of the terms "practices" and "methods" departs from that of Sullivan and Porter. To use de Certeau again, I see "method" as a strategic form of organizing research activities; I see practices as more tactical. The line between the two moves depending on the situation and the nature of the research relations in that situation, most particularly in terms of who is acting (or capable of acting) strategically or tactically.

In terms of Sullivan and Porter's methodological framework, only the concept of practice helps researchers deal with the differences in location and audience—and therefore agency—implied by these examples. Ideological arguments are important, and methods exert a powerful influence. But the practice of research seems most relevant. The issue of practice is often overlooked in writing about methodology. Herndl and Nahrwold frame their argument for critical research in terms of practices, but I read their article as much more about ideology than actual practice. Sullivan and Porter create significant space for research practices, but they don't elaborate much. And again, ideology and method are much more common ways of talking about research. Yet I think research practices may be the most critical, potentially transformative, and as I will show, culturally rich and problematic issue for researchers in technical and professional writing.

To situate my discussion, let me focus on community-based research methodology, which already addresses, in one fashion, the question of research location and which has been useful to me as I have imagined and conducted my own community-based work. Community-based projects, then, serve as my example of where technical and professional writing researchers might locate studies of communicative labor in everyday life or in social institutions other than "the corporation."[7] Community-based research, at least in the ways that community-based researchers talk about it, does not differ much, if at all, from ways in which writing researchers think about their own work.

- There is a concern with power and ethics: why research is being conducted, by whom, in whose interests, and to what ends.
- There is a concern with representation and identity in both the research and writing processes.
- There is vigilance and self-reflection in terms of understanding what drives research: problems, people, commitments, questions, and more traditional disciplinary concerns.

At the levels of ideology and method, it would be difficult to argue that community-based research represents anything new to writing researchers.

When we think about methodology in terms of research practices, however, meaningful differences driven by research location begin to emerge. But there is no established literature in rhetoric and writing detailing research practices, and so here I present a set of practices that have been important to me or that are discussed in the work of community-based researchers and applied anthropologists (two communities of researchers who wrestle with issues of location, audience, and the impact of their research). These categories are not new to research methodology; they are, however, sometimes invisible given the ways we talk about methodology:

1. *Initiation*, or where do studies come from?
2. *Access*, or how do I get permission to do my work?
3. *Participation* (with sponsors, clients, all those impacted; in planning, design, method, analysis, and communication)[8]
4. *Studying up* (or studying one's client or sponsor as well)
5. *Local politics* (mediation, advocacy, relationship building and maintenance, community and political mapping)[9]
6. *Communication* (as a day-to-day research practice itself as well as in myriad settings during the research process regarding "non-research issues"—all in addition to the communication of the research results themselves)
7. *Sustainability*

At this point, I want to pull out a couple of these areas of practice and complicate them in order to show how these practices need to be thought of differently in community contexts, and by extension, how research practices might become a more conscious "space" for research agency. The first concerns access. In some methodological discussions, access is talked about as a process of "getting in," of gaining entry to a site in order to carry out one's study (e.g., Hammersley and Atkinson; Van Mannen; Adler and Adler; Warren; chapters in Rylko-Bauer, van Willigen, and McElroy). Nearly all of the discussion centers on the linked issues of gatekeeping and sponsors, particularly the issue of how a gatekeeper becomes a sponsor. These discussions of access are often most interesting when they focus on the importance of personal relationships and networks for gaining access, for this is often the route for many researchers.

Very little in these discussions of access actually focuses on the practices of entry, or how one actually goes about negotiating access: the ethical and political rules of the situation; the moment-to-moment construction of relationships (Adler and Adler as well as Warren are more fine-grained). It is easy to see access strategies such as leaning on personal or work contacts, persuading a gatekeeper, or deception as manipulative. Furthermore, how exactly does one construct a personal network, especially in institutional contexts where one isn't connected already? As a practice, access in community-based work might best be thought of as bargaining because as researchers, we often assume that our work will be useful on our own terms, when in fact our community partners may see our usefulness differently (see Grindstaff on bargaining). We need to come to some understanding, then, of usefulness. Ruth Ray and Ellen Barton, for example, have questioned the relevance of their interests, training, and disciplinary homes to the communities they studied. They have also noted the ways in which participants in their studies turned the research—mostly the practices and processes—into something that made sense to them and their own needs. How participants make a project meaningful to them is one type of usefulness. But there is a deeper and more activist notion of usefulness at play here, and that is conceiving, designing, and conducting studies with community partners that further their long-term goals in sustainable ways. The practice of critical research in institutional contexts is promising in this regard, but only if usefulness can become a primary epistemological, ethical, and political value. Communities (individuals, organizations, institutions) are sophisticated, busy, and at times suspicious. In each case, we need to negotiate the terms of our access, and therefore, the very design of our studies.[10]

The various practices of access we might imagine are largely invisible in technical and professional writing research, and when discussed in passing as part of larger research methods, are often stripped of their cultural, rhetorical, and epistemological weight. In the risk communication project I am currently working on, access is *the* critical issue and was a sticking point for the better part of a year. In fact, negotiating access required that we develop cultural understandings, adapt our interpersonal skills on the fly, and negotiate the very nature of the project. As we continue to do this work, we are participating in the cultural construction of the community we are working in (and with whom we are working). We have their narratives; we have historical documents; and we have our own narratives (field notes) that we use to come to some tentative understanding of what is going on. We are also laying the groundwork for any future intervention in the risk situation at hand, as the terms of access will shape both the spaces where agency is possible and the rhetorical tools we might develop with those with whom we are working. The community groups we are working with finally began to trust us after a year of this work, a relationship we solidified over coffee at the kitchen table. All this, I want to suggest, is a function of our *practices* as researchers, the productive activities that we and those with whom we are working engage in on a day-to-day basis within the framework of a research project.

The second practice I want to complicate is communication. One of the issues that is important to applied anthropologists and that is connected to the usefulness of their work is when, how, and with whom research results are communicated. Some anthropologists argue for never producing a final report; they argue for preliminary or process reports delivered differently to different stakeholders at different moments—sometimes orally, sometimes in writing, sometimes in meetings, and sometimes in policy documents. Of importance here is the realization that the researcher doesn't control many of these communicative moments. The rhythms of the project, community needs, decision-making timelines, and other such things drive the moments in which this sharing must take place. Furthermore, because research must be communicated to various audiences, study design must be, at least in part, driven by those audiences. For example, if certain types of quantitative results are most persuasive to an audience, then the design ought to attempt to account for this and provide them. Similarly, if there are important civic or community meetings that are planned or materialize, then the ability to quickly produce artifacts for those audiences must be anticipated. This impacts data collection, data analysis, and how one writes up the research. In a word, the need to communicate on-the-fly in community-based work changes the nature of the research, and so demands a type of flexibility that is *practice* driven.

To this point, I have tried to make visible research practices that often remain invisible in how we write about research and to suggest the deep cultural issues that are embedded in these practices. A final example, then, to show them as a source for agency. The risk communication project I have mentioned is located in the community of Harbor, a city that has as much industrial density as any area in North America. In Harbor, there exists a short river channel that links various industrial operations with the lake. Periodically, this channel must be dredged to allow for barge traffic. Given the industrial density in Harbor over a prolonged period of time, the sediments in the channel are heavily polluted. Currently, these sediments flow into the lake, polluting that water body. Thus for navigational and environmental reasons, the channel must be dredged. Dredging these sediments, however, creates another set of problems, as the dredging operation threatens to resuspend contaminants in the water. Furthermore, the transportation of the sediments creates risks, as does the disposal and treatment of the sediments—currently planned to take place in an open confined disposal facility (essentially a landfill protected by clay walls). The project is planned for thirty years, so it is a project of some size. The confined disposal facility will remain open—that is, uncapped—for those thirty years, meaning that there is also a risk of air pollution due to blowing dust particles. To make matters more difficult, the confined disposal facility is located within a few hundred yards of two schools, and the community itself is largely Hispanic, African American, and working class.

Needless to say, this is a project of considerable complexity. Currently, there are two federal agencies, one state agency, two local governments (with their various management and technical functions), four universities, and a number of community-based organizations (some fluid, some stable) involved in deliberations regarding the project. The project touches on issues of engineering (civil, chemical, and environmental), dredging technologies and operations, public health assessment, airborne contaminant research, geology, and a host of legal, procedural, and ethical issues. The citizen groups participating in the decision making processes associated with this project are at a considerable disadvantage, particularly given the fact that they do not have the resources to hire their own experts. Despite the disadvantages, these citizen groups must act if they have any hope of directing the course of the deliberative process, and to engage effectively, they must create new knowledge about the issues at stake and about the community and be persuasive in their presentation of this knowledge.

As part of this project, my colleague, Stuart Blythe, and I work as a communications experts for Technical Outreach Services for Communities (TOSC), an outreach initiative located in Civil and Environmental Engineering at Michigan State University. The presence of communication experts on this team is a function of the level of mistrust that exists in the community and the lack of meaningful and effective communication within the community. Citizens do not trust either their local government or the federal agencies involved. Both government and other agencies find the citizens irrational and unscientific (a commonplace of risk situations). Nobody fully trusts TOSC. TOSC's purpose is to provide reviews of the science and engineering of this project in the interests of the community (this means that community organizations provide questions and concerns and TOSC leverages expertise to address them). But how can TOSC be effective in a situation of such mistrust and miscommunication?

The communication research focus of this project is fairly straightforward. We are concerned with three questions: (1) Who/what is the community? (2) How does it characterize and understand the project (including understanding how community groups "do science")? (3) What are the effective communication practices in this community? To this point, however, it has been the research practices that have been most transformative. I have already mentioned the practices of access that took nearly a year to unfold; again, for nearly a year, all TOSC did, through us, was work on relationship building, which was highly unusual for TOSC. However, we have achieved something. Our most transformative move to this point—and the "our" here refers to the communication researchers in collaboration with one community group—was to change how technical information is communicated to the community.

The standard practice for communicating technical and scientific information to the community is the public meeting. All expert agencies associated with this project, including TOSC, use roughly the same method. Technical information is produced in the form of a report, and that information is reported to the community at the meeting with time for questions and answers. For various reasons, this practice has failed in Harbor. The ways it has been implemented in the past are widely perceived to be manipulative, and besides, highly technical and scientific information is difficult to absorb in a thirty- to forty-minute presentation. It is even more challenging to ask tough questions, and it is impossible given the time frame to challenge the science with contrary science or interpretations.

At our urging, TOSC has developed a new model. Once the technical report is produced (and we have changed how these reports are written and designed, including a community review before publication), it is passed around the community weeks before the meeting. Some of the communication channels used are those networks already in place and utilized by community organizations. Then, after a couple of weeks, there is a public meeting that begins with a short gloss of the report and a much longer time for questions and discussion. This model allows people time to read, to consult with each other and with their own network of experts, and to come to the meeting prepared. So far, this model as worked well in three senses: (1) community organizations like this way of getting information, (2) they have experienced the meetings as less manipulative and frustrating, and as a result (3) they have given TOSC the benefit of the doubt. They listen to us, and in some cases, have started to trust that we are not there to hurt them.

The example of changed reporting practices is important because these practices are in some sense distinct from the research project. That is, we could have conducted our research without arguing for changed reporting practices, and we certainly could write about this research without mentioning them. In addition, neither ideology nor method enables these changed practices—at least not without the situated activities of research practice. In this case, the activities that lead to changed reporting methods cross the categories of practice mentioned earlier but certainly involve issues of participation, communication, attention to local politics, and my favorite, "studying up" (in collaboration with one community organization, we placed TOSC itself under the microscope). The agency within this research project, at least to this point, has been a function of the practice of research. In order for research practices to be ethical and transformative, however, we must pay attention to them. Practices, and the relationships in which they are imbricated, can just as easily be dominating and limiting.

It is the ambivalence of practices, even within a project that is shaped by "correct" ideologies and methods, that makes, cultural and rhetorical intersections necessary for research methodology in technical and professional writing. Blake Scott, for example, examines HIV testing practices and argues that they need to be understood within an "ensemble of heterogeneous cultural practices" (229). And so he takes his readers through an examination of biomedical and public policy arguments and a substantive body of rhetorical theory. All are necessary to begin to understand the technoscience of HIV testing. But the most interesting moment is when Scott utilizes these new understandings to help rewrite test counseling materials and a risk questionnaire. Scott is able to locate—to operationalize—his cultural critique within the day-to-day discursive practices of specific institutions. He is able to show the implications of his critique by illustrating how a flawed cultural rhetoric of risk is embedded in the questionnaire. Far more important, the practice of his research methodology put him in a location/position, allowed him to see and understand professional writing as important, and created the space for rhetorical agency. Scott's "social factory" is a counseling (and testing) organization. Within this institution, culture is not background; the cultural moment is actively created through his research, and it is only this rhetoric of culture that opens up space for the seemingly mundane intervention of revised documentation. Only such cultural and technical understanding makes a moment of practice such as this available and meaningful.

My point is to show that location matters and to point out as well that a methodological framework such as the one Sullivan and Porter outline might provide the heuristic tools necessary to imagine professional and technical writing research that works in the problem space of communicative labor within a wide range of social institutions. With respect to practices themselves, however, there is really something deeper at stake: these practices are an important way in which research relations are constituted, *and* they are knowledge-producing. They do in fact lead to new and different understandings of a project and should be understood to have epistemological value, not just procedural utility. The knowledge produced from these practices certainly shapes study design, which in turn shapes the results of the larger inquiry, but they lead to community understandings and personal relationships that are valuable in their own right. Furthermore, practices are one way to locate agency and therefore the possibilities for critical and transformative research. In technical and professional writing, we have a critical discourse regarding research (ideologies) and we have incorporated methods that make sense given the range of inquiry concerns in the field. We do not understand practices as well as we might, and therefore, I think, we do not have a way of coming to grips with agency at the research scene. To do so, however, requires more than a concrete sense of research practices; it requires ways of understanding cultural production.

Research, then, produces culture and it produces its own possibilities for change. More important, critical research in technical and professional writing helps others be productive in their day-to-day lives, their communities, and in the institutional spaces that shape their lives (and that we hope they, in turn, can shape). The critical research tradition in technical and professional writing is important. Critical research could be strengthened by rhetorical studies of culture and could therefore move technical and professional writing research into the study of "rhetorics of the everyday" or writing in the social factories around us, spaces where the field can make a significant contribution to public life.

Notes

1. An immediate caveat: Much of the discussion in this chapter will focus on researcher agency, but when I discuss research practices later in the chapter, I see these activities as opportunities for participant agency. In both cases, agency is limited to the institutional scene of a research project. The scope for agency is clearly limited in my discussion, but whether or not the forms of agency I will discuss are *limiting* is an open question.

2. Herndl and Nahrwold create a bell curve in which most of the research is descriptive (the middle), while significantly less supports instrumental workplace practices or critiques them (the ends).

3. What is not adequately accounted for in my discussion is the way in which research participants shape the research scene as well. This issue is fundamental, and my exclusion of it is motivated by focus concerns, not any sense of relative importance.

4. Having said this, it is certainly true that intellectual work can have a transformative effect within the academy, changing how we teach and professionalize students. The claims of cultural studies are much broader. I *am* claiming, however, that production is potentially more transformative than reception. Though the lines between the two are messy and ultimately untenable, it is possible and necessary to draw distinctions between types of research production. An essay is not the same thing as a grant proposal. Neither are the same thing as any number of other activities, symbolic and otherwise, that take place within a research project. Which is more "productive" or "transformative" depends on the situation, of course. However easy it is to deconstruct binaries such as reception-production, it seems important to do the difficult work of drawing distinctions between types of productivity as forms of rhetorical agency.

5. We might ask, for example, about the class positions of technical writers and how these positions matter. Or about the status of this type of communicative labor within organizations and with respect to the dynamics of globalization.

6. There are still deeper connections to be made here. Janet Atwill argues that rhetoric is a productive art—that is, that it utilizes/produces a form of productive

knowledge. This allies rhetoric with arts of making (e.g., medicine and architecture), and in Atwill's schema, also makes rhetoric a transgressive art that challenges or redefines relations of power (7).

7. I do not mean to suggest that "the community" is a single location or concept. The term glosses a wide range of institutional sites.

8. Of note here is the importance of relationship building and maintenance, which is an issue that crosses each of these categories of practice. It is also worth mentioning here that the participation required of community-based research also includes collaborating on community events and projects that are genuinely not connected to a particular research project.

9. Research practices associated with local politics are complex. A researcher needs to learn to listen carefully and read between the lines. One needs to learn local rhythms, or when to move slowly and when to push. Researchers need to understand that one is always already a participant, and that one's choices have concrete political implications.

10. But do researchers in technical and professional writing discuss or teach how to negotiate? Do we know when or where this practice might or should take place? Is it over the phone? In person? At a social event or during "work" hours? And how often do we know this early in a project? And finally, what is the offer (and there is always an offer, some articulation of the "good" of the research)? How *is* the community, if we can think of identity here in the singular, going to benefit, always being mindful that the benefit a community partner might attach to a research project might be radically different from the benefit a researcher anticipates; being mindful as well of how arrogant we can appear in announcing how our work will help others. In my experience, this bargaining process is critical to access, but it would be a mistake to see this as mere horse-trading. If approached as a research practice, this process of negotiation is really part of study design and has significant epistemological value as well.

WORKS CITED

Adler, Patricia A. and Adler, Peter. *Membership Roles in Field Research.* Newbury Park, CA: SAGE, 1987.

Atwill, Janet M. *Rhetoric Reclaimed: Aristotle and the Liberal Arts Tradition.* Ithaca, NY: Cornell UP, 1998.

Barton, Ellen. "More Methodological Matters: Against Negative Argumentation." *College Composition and Communication* 51 (2000): 399–416.

Berlin, James A. *Rhetorics, Poetics, and Cultures: Refiguring College English Studies.* West Lafayette, IN: Parlor Press, 2003.

Blyler, Nancy. "Taking the Political Turn: The Critical Perspective and Research in Professional Communication." *Technical Communication Quarterly* 7 (1998): 33–52.

Charney, Davida. "Empiricism Is Not a Four-Letter Word." *College Composition and Communication* 47 (1996): 567–593.

———. "From Logocentrism to Ethnocentrism: Historicizing Critiques of Writing Research." *Technical Communication Quarterly* 7 (1998): 9–32.

Cintron, Ralph. *Angels' Town: Chero Ways, Gang Life, and Rhetorics of the Everyday.* Boston: Beacon Press, 1997.

———. "'Gates Locked' and the Violence of Fixation." *Toward a Rhetoric of Everyday Life: New Directions in Research on Writing, Text, and Discourse.* Ed. Marin Nystrand and John Duffy. Madison, WI: U of Wisconsin P, 2003. 5–37.

Cooper, Marilyn M. "Distinguishing Critical and Post-Positivist Research." *College Composition and Communication* 48 (1997): 556–561.

de Certeau, Michel. *The Practice of Everyday Life.* Berkeley: U of California P, 1988.

Doheny-Farina, Stephen. "Research as Rhetoric: Confronting the Methodological and Ethical Problems of Research on Writing in Nonacademic Settings." *Writing in the Workplace: New Research Perspectives.* Ed. Rachel Spilka. Carbondale, IL: Southern Illinois UP, 1993. 253–267.

Foley, Douglas. "Does the Working Class Have a Culture in the Anthropological Sense?" *Cultural Anthropology* 4 (1989): 137–163.

Grindstaff, Laura. *The Money Shot: Trash, Class, and the Making of Talk TV.* Chicago: U of Chicago P, 2002.

Hall, Stuart. "Cultural Studies: Two Paradigms." *Media, Culture, and Society* 2 (1980): 57–72.

———. "The Emergence of Cultural Studies and the Crisis of the Humanities." *October* 53 (1990): 11–90.

Hammersley, M., and Atkinson, P. *Ethnography: Principles in Practice.* 2nd ed. London: Routledge, 1995.

Herndl, Carl G. "Writing Ethnography: Representation, Rhetoric and Institutional Practices." *College English* 53 (1991): 320–332.

———. "Teaching Discourse and Reproducing Culture: A Critique of Professional and Nonacademic Writing." *College Composition and Communication* 44 (1993): 349–363.

———. "The Transformation of Critical Ethnography into Pedagogy, or the Vicissitudes of Traveling Theory." *Nonacademic Writing: Social Theory and Technology.* Ed. Ann Hill Duin and Craig J. Hansen. Mahwah, NJ: Lawrence Erlbaum, 1996. 17–34.

Herndl, Carl G. and Nahrwold, Cynthia A. "Research as Social Practice: A Case Study of Research on Technical and Professional Communication." *Written Communication* 17 (2000): 258–296.

Lindquist, Julie. *A Place to Stand: Politics and Persuasion in a Working Class Bar.* New York: Oxford UP, 2002.

Longo, Bernadette. "An Approach for Applying Cultural Study Theory to Technical Writing Research." *Technical Communication Quarterly* 7 (1998): 53–74.

Nelson, Cary, Treichler, Paula A., and Grossberg, Lawrence. "Cultural Studies: An Introduction." *Cultural Studies.* Ed. Lawrence Grossberg, Cary Nelson, and Paula A. Treichler. New York: Routledge, 1992.

Porter, James E., Patricia Sullivan, Stuart Blythe, Jeffrey T. Grabill, and Libby Miles. "Institutional Critique: A Rhetorical Methodology for Change." *College Composition and Communication* 51 (2000): 610–642

Ray, Ruth, and Barton, Ellen. "Farther Afield: Rethinking the Contributions of Research." *Under Construction: Working at the Intersections of Composition Theory, Research, and Practice.* Ed. C. Farris and C. Anson. Logan: Utah State UP, 1998.

Scott, J. Blake. *Risky Rhetoric: AIDS and the Cultural Practices of HIV Testing.* Carbondale: Southern Illinois UP, 2003.

Selfe, Cynthia L., and Richard J. Selfe, Jr. "Writing as Democratic Social Action in a Technological World: Politicizing and Inhabiting Virtual Landscapes." *Nonacademic Writing: Social Theory and Technology.* Ed. Ann Hill Duin and Craig J. Hansen. Mahwah, NJ: Lawrence Erlbaum, 1996. 325–358.

Stewart, Kathleen. *A Space at the Side of the Road: Cultural Poetics in an "Other" America.* Princeton, NJ: Princeton UP, 1996.

Sullivan, Patricia, and Porter, James E. *Opening Spaces: Writing Technologies and Critical Research Practices.* Greenwich, CT: Ablex and Computers and Composition, 1997.

Van Mannen, John. *Tales of the Field: On Writing Ethnography.* Chicago: U of Chicago P, 1988.

van Willigen, J., Rylko-Bauer, B., and McElroy, A., eds. *Making Our Research Useful: Case Studies in the Utilization of Anthropological Knowledge.* Boulder: Westview, 1989.

Warren, Carol A. B. *Gender Issues in Field Research.* Newbury Park, CA: Sage, 1988.

Chapter Seven

Living Documents

Liability versus the Need to Archive, or Why (Sometimes) History Should Be Expunged

Beverly Sauer

Cultural studies examines the emergence and transformation of discourse within cultural and historical contexts, critiques the functions of discourse in power-laden processes of knowledge creation and legitimation, and reflexively accounts for the analyst's role in these processes (Longo). Cultural studies is thus well situated to help technical communicators understand both the culturally specific origins of particular practices and the continuing regulation of culturally constituted practices in new and emerging contexts. Further, cultural studies, with its tradition of critiquing the politics and ethics of capitalist institutions and material conditions, can provide a theoretical ground for investigating technical communication in difficult cross-cultural and material contexts such as mining. In South Africa, for example, communication practices may continue to reflect the institutional structures of (capitalist) mines under apartheid despite changes in the official regime (Dubow). Cultural studies can thus help technical communicators understand the historical and institutional origins of communication practices that influence the outcomes of training programs designed to transform the institutional and economic status of workers (cf, Latour, Woolgar, and Salk; Selzer).

A nuanced cultural analysis of such contexts challenges technical communicators to rethink how they apply theoretical constructs in real-time worksites, however. Although technical communicators can use cultural theory to critique scientific positivism or rationality and develop a "postpositivist" science, technical communicators must still understand how positivist values "become 'significant'" (Brown 16) both historically and politically within institutions where research is intended to improve the health and safety of workers. In demonstrating how technical standards erase the uncertainty of work in complex and dynamic material sites, for example, technical communicators must still acknowledge the efficacy of standards in hazardous environments where labor unions have been most active in the pursuit of regulatory reform and safety regulation. At the same time, in looking at standards as historical artifacts, we must remember that those standards may never have been truly "modern," that is, separable from their institutional and cultural contexts that give them meaning (Latour *We*).

This chapter examines the notion of standards as "living documents"—documents that are continually revised and updated in response to new information, new technologies, and changing social and institutional practices. The present chapter describes the process of regulatory revision and review of one large government agency in order to articulate (1) the tension between science and the social in local risk decision-making; (2) the challenges of developing web-based documentation systems as a response to increasing technical and social complexity; and (3) the problem of maintaining document archives for historical analysis versus the need to expunge outdated and potentially dated standards.

After discussing the role of cultural analysis in technical communication research, this chapter applies the methods of cultural theorists to examine, reflexively, whether the complexities of technical documentation in hazardous environments can also inform cultural historians' understanding of the assumptions they apply to interpret other types of discourse.

SOCIOCULTURAL STUDIES OF SCIENCE AND TECHNICAL COMMUNICATION

Because cultural theorists interrogate the economic and political dimensions of science, cultural theory provides an almost ready-made critique of regulatory practices within hazardous industries whose structures reflect their origins in early capitalist notions of power.[1] Social science theorists have used cultural theory to examine (and critique) the social construction of scientific knowledge by a network of human and nonhuman actors (Knorr; Latour, Woolgar, and Salk; Lynch and Woolgar; Sismondo). Some studies have demonstrated the role of local knowledge in risk decision-making and argued for a more inclusive notion of what counts as science, particularly in regard to the public understanding of risk (cf. Irwin; Irwin and Wynne).

More recently, Latour has questioned the adequacy of social-theoretical explanations of science as a process of "rationalization" ("Re-modernization" 46). Responding to Ulrich Beck's argument that "re-modernization" is occurring, Latour argues that the control presupposed in the notion of the positivist/modernist project was always a "fiction," despite modernist efforts to control "externalities" (36–37). He proposes a counternarrative of modernist scientists who, in seeking to distance themselves from social context, find themselves inadvertently more entangled in the social world—like the scientists in the atom bomb project who claimed to be just doing science even as they reshaped the modern social world. In exposing this "fiction," Latour warns social science thinkers to resist creating their own master narratives when they seek to find an overarching theory of science and its representations.

Because they work from broad theoretical assumptions about the production of scientific knowledge and discourse, social scientists have sometimes not addressed practical questions about the implementation of theory in practice. Designing information systems that support risk decision-making and hazard assessment in local sites, for example, entails practical questions about competing models of risk assessment, the continual adjustment of standards based on observation and experience, and the demands that new technologies place on existing standards. Given the complexity of documentation, can we use social science theory to build documentation systems that simultaneously facilitate risk decision-making and capture the dynamism and contingency of socially constructed knowledge? What role should social science theorists play in creating standards that protect real-life human agents? Is there a role for the social epistemologist in the everyday work of underground coal miners?

Historians working within a social-theoretical model have helped us understand the effects of organizational communication on risk assessment and management (see Clarke and Short). They have revealed the limits of written documentation (Olson; Sauer "Sexual," "Embodied Knowledge"); the emergence of genres within institutions and disciplines (Berkenkotter and Huckin; Bazerman); and the processes by which risk is "normalized" in large bureaucratic institutions (Vaughan, "Dark," "Rational," "Trickle-down," "Challenger"). In her reconstruction of the *Challenger* launch decision, for example, Vaughan describes a "pattern in which signals of potential danger—information that the booster joints were not operating as predicted—were repeatedly normalized in engineering risk assessments prior to 1986" ("Challenger" xiii). She describes how NASA continually devalued the subjective observations of engineers; discounted "engineering feel" in the absence of (hard) scientific data; and failed to act because "acceptable technical arguments in FRR had to meet the quantitative standards of scientific positivism" (354; see also 221–222). Vaughan ultimately blames "the cultural emphasis on scientific positivism and quantitative arguments [that] systematically excluded nascent engineering theories" (263). Her work provides valuable context for more specific rhetorical studies of communication at the moment of disaster (Dombrowski; Herndl, Fennell, and Miller; Winsor). But Vaughan remains cautious about recommending "subjective experience" as a warrant for risk decisions, particularly when "structural changes in a system can also backfire": "What happens in flight-readiness decisions if the intuitive and subjective are acceptable in risk assessments? What happens if bureaucratic mandates are less rigorous? How can we truly explore the consequences of cultural tinkering in advance of such changes in order to minimize the ironies of social control?" (419)

My own analyses of documentation practices in the coal mining industry have combined cultural analysis (at the systems level) with rhetorical and linguistic analysis in order to examine the representation of local knowledge in large regulatory agencies.[2] I have documented how individual documents fail to capture the embodied experience and observations of individuals whose knowledge might contribute to safer practices. In arguing for a more nuanced understanding of the role of embodied experience in everyday risk decision-making, I also have warned readers to be cautious in asserting the value of experience as the sole warrant for risk decision-making (Sauer "Embodied Knowledge," *Rhetoric*). If we valorize individual observation and experience, I have argued, we also place the burden of responsibility on individual actors who must personally bear the responsibility for deciding whether the putty or burn-through that they observe is qualitatively worse (or better) than it appeared in previous accounts of problems. Critiquing the (necessarily) reductive character of risk regulation may also shift the liability for risk assessment from management to line workers who do not have the equipment or authority to protect themselves and colleagues against the (sometimes arbitrary) practices of unscrupulous inspectors—particularly in countries like South Africa, which have not developed the regulatory standards that protect U.S. workers (Sauer "Language"; cf. Moodie and Ndatshe).

As Curran argues, moreover, standards can become "dead laws for dead men" if agencies fail to implement and enforce standards in practice. Writing in the midst of the Reagan administration's moratorium on regulation throughout the federal system, Curran describes how "legislative rhetoric can address a crucial issue in such a way as to satisfy concerns about safety, and simultaneously allow sufficient flexibility for necessary reinterpretations later" (129). While Curran is cautious in arguing against a deliberate plot between agencies and coal-mining operators, his analysis demonstrates that the most carefully written safety regulations ultimately depend on interpretations influenced by social, economic, and institutional conditions beyond the control of the writer's "original intent."

With the publication of standards on the Internet, regulators now confront even more opportunities for increasing the interpretative flexibility of standards in new contexts, but there is also a danger in the proliferation of standards without regulatory control or oversight. The following discussion of standards as "living documents" challenges technical communicators to discover new methods for documenting the multimodal and heterogeneous character of knowledge critical to risk management and assessment at the same time that it cautions technical communicators to be careful in assessing popular knowledge in light of its potential hazard to the community.

"Living Documents" Within a Regulatory Framework

Historians use the term "living document" to refer to the dynamic and continual evolution of meaning in documents like the Constitution or Bill of Rights. In this sense, documents continue to "live" (e.g., have meaning) in new contexts, and new interpretations and new contexts give life to framers' original intentions. When meanings are ambiguous or unclear, legal theorists may attempt to recover the framers' original intent or challenge existing meanings with new and often creative interpretations of the literal text.

Noah demonstrates the importance of archival histories in the interpretation of regulatory intent. As he explains, legal interpretations of regulatory intent are often dependent on meanings encoded in postpromulgation documents that are vulnerable to change as administrations change or as agency executives revise their stance based on different regulations. Noah contrasts the relative scarcity of prepromulgation materials in earlier regulatory rulemaking with the quantity and accessibility of prepromulgation documents over the past thirty years, concluding that "courts should embrace such valuable interpretive materials [as evidence of regulatory intent] rather than rushing to defer to the dynamic interpretation that an incumbent administration finds most expedient" (1).

Although Noah asks courts to draw on prepromulgation documents as evidence, his conservative interpretive methods do not take into account the problem of interpreting standards in new and emerging contexts that were not (and could not be) envisioned within the necessarily limited scope of the framers' original intent. In some contexts, not expunging the archival record could present workers with conflicting rules and may compromise their health and safety.

Cultural theorists can help technical communicators design living documents that preserve the historical archive without compromising safety. Technical communicators might begin with questions about the nature of documentation practices and their relation to cultural history, power, and material, embodied experience: Who constructs standards for documentation? How is documentation shared? Who makes judgments about the relevance of shared information? Who has access to documents in the public and private domain? What is the impact of poor documentation practices on the institutional and social dimensions of risk? What kinds of information do we need to capture to prevent disasters? What historically specific assumptions inform these documentation processes? How do these documentation processes continue to influence practices in hazardous environments?

Previous studies of large- and small-scale disasters have focused on one-time events like the *Challenger* disaster or events at Three Mile Island rather than the quotidian problems of regulating hazardous environments. Technical communication textbooks discuss problems of instruction, but until recently, they have paid little attention to the contexts in which instructions are applied, the uncertainties of risk prediction in hazardous environments, the pressures that workers face as companies look for cheaper (off-shore) sources of labor, or the difficulties of interpreting generalized regulations and instructions across multiple (and often diverse) worksites.

The notion of living documents illustrates how the material uncertainty of the workplace is instantiated in regulatory practice.[3] When Congress enacts safety legislation, the framers of the regulation propose a *regulatory framework* that can change in response to the specific needs of management and labor. This regulatory framework functions as a general outline of the components of safety regulation in particular industries and includes topics like mandatory health and safety standards; inspection standards; authorization of appropriation; mandatory health and safety training; inspections, investigations, and record-keeping; injunctions; and penalties.

Despite the apparent certainty embodied in the notion of legal standards, regulatory frameworks invisibly embody notions of uncertainty, change, and unpredictability. For every Act, Congress first defines the structure of the Act in the initial legislation. This structure remains the same (with a few exceptions) for the duration of the Act, but the specific fines, safety training standards, and record-keeping requirements change as Congress enacts stricter or more lenient enforcement rules. Thus, the current (when this article was written) Federal mine safety act is officially titled Federal Mine Safety and Health Act of 1977, Public Law 91-173, as amended by Public Law 95-164 (U.S. Dept. of Labor). The Act warns readers not to use the document as a legal citation because the Act (as presented online or in paper version) may reflect outdated and thus dangerous provisions:

> This document was prepared within the Office of the Solicitor, Division of Mine Health and Safety, Department of the Interior, and reflects changes to the Federal Coal Mine Health and Safety Act of 1969 [Pub. L. 91-173]. The changes are based upon the Federal Mine Safety and Health Amendments Act of 1977 [Pub. L. 95-164]. Do not use this document as a legal citation to authority. (U.S. Dept. of Labor, 2003)

Although we can speak of the specific provisions of the original 1969 Mine Act (or its 1977 revision) as enacted by Congress, many of the Act's original standards have been revised and updated to reflect new research, new policy, or a better understanding of mine safety. Agency personnel often refer to the Mine Act (or simply "the Act") in general terms without reference to a specific year or date. Outdated provisions are quickly excised from the living record; they are hard to find except in archives. The structure of the Mine Act reserves specific sections for specific kinds of regulation, but the details of regulatory practice change over time. The standards thus become a "living document" that is constantly reviewed and revised within the parameters of the original scope and intent of the authorizing legislation.

The Act is a "living document" in a second sense to the extent that agencies and institutions continually interpret and reinterpret the provisions of the act in specific contexts. Most legislation acknowledges that no single standard can be generalized across diverse environments and mining operations. Although the Mine Act of 1977 forbids any change or reinterpretation that diminishes worker safety, management or workers can petition to modify particular standards if they can demonstrate that the new practice will not diminish the health and safety of workers [§101 (c)]. Regulators also recognize that local conditions ultimately dictate safe practice.

When accidents occur, agencies reexamine standards. Much of the discussion surrounding standards demonstrates the tension between the need to develop enforceable standards that reduce uncertainty and increase safety. Industry and labor unions disagree, for example, about the meaning attached to "imminent danger." Should it include potential but (as yet) nonexistent problems and conditions that might potentially affect the health and safety of workers? Do we need to be 100 percent certain? Fifty percent certain? Is a one-in-three chance of disaster "imminent"? (McAteer and Galloway). The Act is thus a living document in a third sense, as interpreters give meaning to specific standards across multiple and often diverse worksites.

As the following discussion suggests, the politics of this discussion often inverts the political position of most cultural-theoretical critiques of regulatory practice. Cultural theorists, for example, may inadvertently find themselves arguing against safety regulation when they talk about the uncertainty of standards and the difficulties of applying general standards in specific contexts. Marxist theorists may find themselves aligned with management and positioned against labor unions in the debate over so-called positivist attempts to reduce uncertainty.

Cultural studies of technical communication can reveal the underlying contexts that give meaning to notions of certainty and uncertainty at specific moments in history within specific documents and practices. To advance this project, technical communicators must also help cultural theorists understand the difficulties and limitations of written communication in the context of risk.

STANDARDS AS LIVING DOCUMENTS WITHIN LARGE REGULATORY AGENCIES

When coal mine disasters occur, the U.S. Mine Safety and Health Administration investigates the cause or causes of the accident in order to prevent disasters in the future. Much of this debate involves the degree to which standards enabled workers and management to regulate risk in practice in hazardous worksites: Did management's failure to conform to standards cause the disaster? Were standards adequately enforced and monitored? Were standards clearly written? Were workers properly training to comply with standards? Will tighter standards prevent similar disasters in the future?

Although many standards specify strict limits on toxic substances (like the 1 percent limit on methane in coal mines), most standards are not "bright lines" (U.S. Presidential/Congressional Commission). Standards are heterogeneous and complex texts that attempt to formulate in writing the accumulated body of scientific and local knowledge about preventing disaster in hazardous worksites (Sauer *Rhetoric*). Standards reflect reasonable and practicable solutions to complex technical problems in the workplace. But they are also the outcome of many deliberations about the value of human life and the distribution of risk among affected populations. Some mining standards reflect scientific theories about gas behavior, combustion, or rock mechanics [e.g., 30 CFR § 75(301–305)]. Other standards specify general procedures like air sampling without citing specific techniques or codes of practice [30 CFR § 75(301–306)]. Some standards specify testing methods developed by scientific testing institutes like American Society for Testing and Materials (ASTM) [30 CFR § 75(302–303)]. And some specify the sequence of procedures that must occur in preshift examinations and inspections [30 CFR § 75(303)] (Sauer *Rhetoric*).

Some standards "standardize" reasonably practicable procedures that are quickly outdated as new technologies emerge or researchers discover new methods of analysis and testing. Sometimes, standards simply rewrite highly local, commonsense practices. Thus, the miner's warning, "never go under unsupported roof," becomes 30 CFR §75(202) of the Mine Act: (b) No person shall work or travel under unsupported roof unless in accordance with this subpart.[4]

The term "regulation" also has many different meanings in hazardous worksites. At the agency level, regulators construct written regulations that define safety standards and set limits to risky practices and procedures. At the local level, human decision-makers (regulators) work in tandem with mechanical systems (also known as regulators) to monitor and control (regulate) conditions in the environment. To determine whether monitors provide accurate readings, human regulators must continually calibrate and recalibrate mechanical regulators against known standards. Some mechanical regulators automatically shut down production when sensors detect dangerous conditions. But the decision to stop production is by no means automatic. Even when danger is imminent, management and workers work to bring conditions into compliance—estimating risks, predicting outcomes, and implementing precautions to reduce the potential for disaster (Sauer *Rhetoric*).

Standards are continually reinterpreted in an ongoing process of revision and review.

Because safety standards are difficult to write and apply in hazardous environments, Federal agencies continually reinterpret standards in policy memos, compliance guidelines, and program information. This body of interpretation must be written and disseminated so that companies can revise practices and procedures in local contexts. The U.S. Mine Safety and Health Administration sends policy memos to all interested parties as new interpretations arise; the Federal Aviation Administration updates all maintenance standards every ninety days. Outdated standards must be replaced every ninety days; workers are cautioned *not* to rely on memory when they apply standards in the workplace because they may remember outdated standards and procedures.

Regulatory agencies engage in a continual process of regulatory revision and review to keep regulations updated—even when administrative guidelines forbid any new regulation. Although agencies cannot legally reveal the negotiations, attempts at rewording, and multiple comments and interpretations that precede the final draft of accident reports and investigations, congressional overseers can subpoena agency documentation (policy memos and technical reports) in order to investigate problems in the regulatory process.

The sample document in fig. 7.1 demonstrates how individual writers literally "write in" new interpretations. In this example, the layering of responses within the document becomes increasingly more complex as respondents react to comments from previous commentators. As the number of commentators and interpretations increases with each new review, it becomes increasingly difficult to separate writer, respondent, and audience within the text as each commentator raises new questions about the uncertainty of new technologies and the efficacy of standards that were in effect at the time of previous disasters. As the example in fig. 7.1 demonstrates, respondents draw on previous disasters and previous interpretations of regulations to warrant their concerns about new regulations and practices.

In fig. 7.1, Bernard challenges the results of a technical review of the issues raised in previous memos. Bernard's handwritten comments above the typed text make explicit his alternative conception of Utah mining law (478). Bernard draws on a previous accident to challenge the agency's representation that these events were "rare" (483). He argues instead that the same standards were in effect in a previous accident "when MSHA and Co. personel [*sic*] were killed [in] West Virginia" (483).[5] In support of the review, Beason counters that the new ventilation practice has already "proved disasterous [sic]" (461). He bases this assessment on previous accident investigations and recommendations from the American Mining Congress.

Many of the arguments are hypothetical. In one section of this memo, agency writers acknowledged that individuals might encounter hazardous conditions if they entered a mine following an explosion, but these writers assure their audience that previous standards had provided "protection . . . on these rare occasions" (Bernard 490). When Bernard asserts that "it is reasonable to assume that a prudent mine operator" would make smart choices about the location of a fan "absent the regulation" (490), Beason counters: "If all operators were prudent—we wouldn't need any regulations, but history proves that wrong!" (490).

As these exchanges demonstrate, individual documents are not as important as the documented history of risk assessment because reviewers draw on information in previous accounts to assess the cause or causes of new disasters. When accident reports provide inadequate information, agencies and individuals must spend time reviewing and rereviewing documents so that they can reconstruct events and conditions that preceded the disaster. Without adequate evidence, these individuals cannot argue successfully for policies to prevent accidents in the future.

The quality of technical communication affects future risk decision-making in an ongoing cycle of rhetorical transformation within large regulatory industries.

As the previous discussion suggests, agencies depend on adequate and timely technical documentation to interpret and enforce regulations in hazardous environments. This documentation builds a body of experience—a cultural history—that agencies can draw on to help them improve the health and safety of workers. When the cultural history is incomplete, agencies must

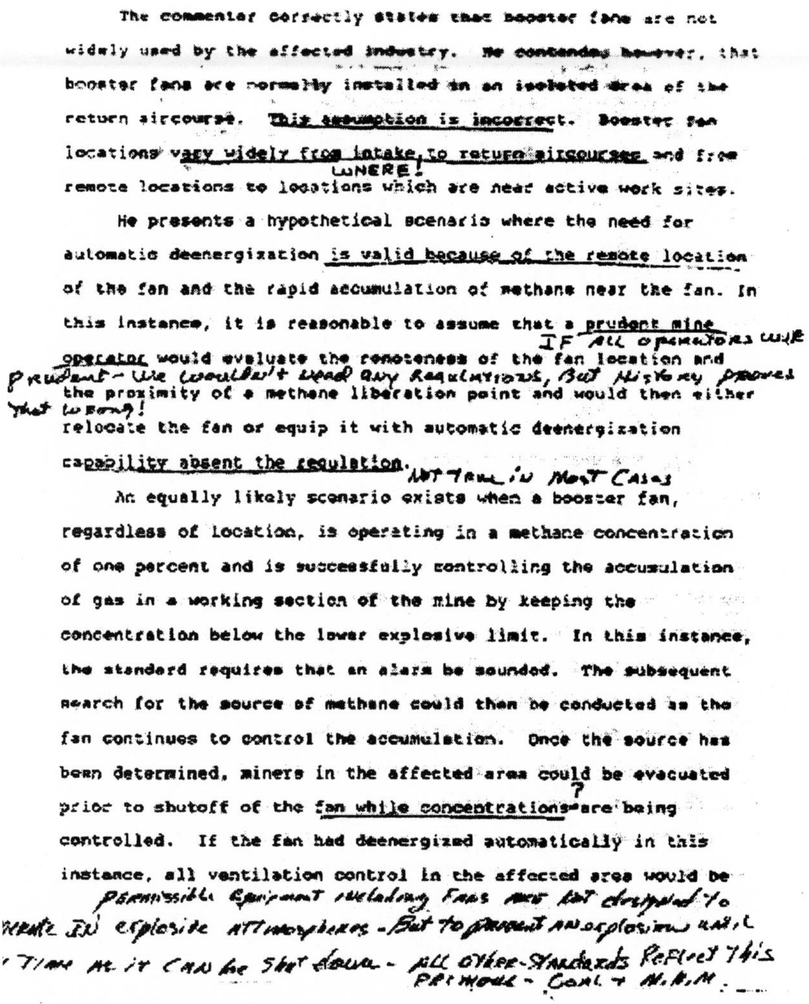

The commenter correctly states that booster fans are not widely used by the affected industry. He contends however, that booster fans are normally installed in an isolated area of the return aircourse. This assumption is incorrect. Booster fan locations vary widely from intake to return aircourses and from remote locations to locations which are near active work sites.

WHERE!

He presents a hypothetical scenario where the need for automatic deenergization is valid because of the remote location of the fan and the rapid accumulation of methane near the fan. In this instance, it is reasonable to assume that a prudent mine operator would evaluate the remoteness of the fan location and the proximity of a methane liberation point and would then either relocate the fan or equip it with automatic deenergization capability absent the regulation.

IF ALL OPERATORS WERE PRUDENT - WE WOULDN'T NEED ANY REGULATIONS, BUT HISTORY PROVES THAT WRONG!

NOT TRUE IN MOST CASES

An equally likely scenario exists when a booster fan, regardless of location, is operating in a methane concentration of one percent and is successfully controlling the accumulation of gas in a working section of the mine by keeping the concentration below the lower explosive limit. In this instance, the standard requires that an alarm be sounded. The subsequent search for the source of methane could then be conducted as the fan continues to control the accumulation. Once the source has been determined, miners in the affected area could be evacuated prior to shutoff of the fan while concentrations are being controlled. If the fan had deenergized automatically in this instance, all ventilation control in the affected area would be

?

PERMISSIBLE EQUIPMENT INCLUDING FANS ARE NOT DESIGNED TO OPERATE IN EXPLOSIVE ATMOSPHERES - BUT TO PREVENT AN EXPLOSION UNTIL TIME AT IT CAN BE SHUT DOWN - ALL OTHER STANDARDS REFLECT THIS PREMISE - COAL + M.N.M.

FIGURE 7.1. Bernard challenges results of technical review. From Bernard, R. L.

reconstruct the history of events from testimony, archival sources, agency reports, and data.

The individuals who produce this documentation are not normally trained as writers, but they write frequently and regularly as part of their work, and their writing must demonstrate that they have adequate and sufficient evidence for their decisions. Unfortunately, few of these individuals have developed the rhetorical strategies they need to describe the relationships known and unknown, the theory and practice, the standards and phenomena they seek to regulate. As a result, administrative courts frequently overturn citations because agency writers have failed to provide adequate and timely warrants for their judgments (Sauer *Rhetoric*).

In the *Rhetoric of Risk*, I have described how documentation practices within agencies often fail to capture knowledge that individuals need in order to understand the material conditions in which they work. The quality of documentation matters because each document serves as a warrant that influences future policy and procedure throughout the industry (Sauer, 2003).

Fig. 7.2 presents a framework for examining the transformation cycle of documentation in large regulatory industries. The Cycle of Documentation in Large Regulatory Industries depicts six predictable moments when individuals have the opportunity to transform information in one form (e.g., embodied sensory perception) for use by new audiences in new contexts (accident analysis and physical modeling): (1) when investigators capture workers' narratives to determine the cause of the accident; (2) when analysts abstract statistics to pinpoint areas that need special attention; (3) when policymakers interpret statistics in light of new technologies and changing political economies; (4) when trainers rethink standards from a miners' perspective; and (5) when operators and management interpret standards and procedures to manage hazards in local sites (cf. Sauer *Rhetoric*).

While the process of rhetorical transformation may be most visible in the so-called dumbing down of documents for lay audiences, rhetorical transformation also occurs when engineers and scientists calibrate local expe-

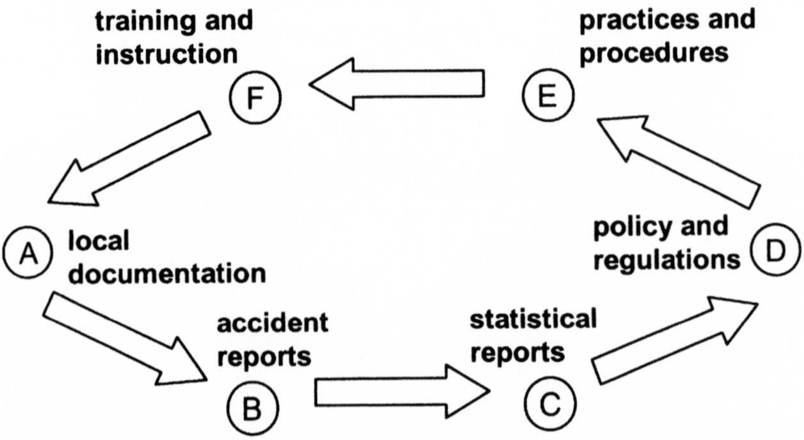

FIGURE 7.2. The Cycle of Technical Documentation in Large Regulatory Industries. Originally appeared in *The Rhetoric of Risk* by Beverly Sauer (2003: 84). Reprinted with permission from Lawrence Erlbaum Associates, Inc.

rience with known scientific standards so that the information can be used as warrants for future risk management and assessment. Molinda and Mark's Coal Mine Roof Rater system (CMRR) illustrates how engineering indices help decision makers quantify observations so that the information can be used to regulate risk in large systems. The process is simple: When workers strike exposed rock with a ball-peen hammer, the shape of the resulting indentation is calibrated with a quantitative measure of roof strength. Table 7.1 summarizes the process.

Local term	←Norms of transformation→	Discourse of risk
rebounds	CMRR⇒ ⇐	>103 Mpa (>15,000 psi)
pits		103-55 Mpa (15,000-8,000 psi)

TABLE 7.1

The CMRR Quantifies Experience

At this point in the argument, many technical communicators might be suspicious of attempts to reinvent what cultural theorists may describe as purely positivist attempts at objectivity. As the following discussion suggests, however, the rhetorical transformation of local knowledge into the formal discourse of risk (and vice versa) has important consequences for risk assessment in local sites. Regulations based on objective empirical data and experimentation can reinforce and support workers' subjective experience. Observations of smoke may indicate fire or diesel fumes (Huntley et al.); the difference is critical. We do not close a coal mine every time a miner feels a chill or hears an unexpected noise. As McCloskey concludes, "The alternative to modernism is not irrationalism" (168; cf. Dawes "Irrationality," *Rational*).

Mine operators frequently argue that it is not feasible to use the most technically advanced scientific measurement devices in an underground mine. In some cases, there is simply no nondestructive method of testing. In the Bilsthorpe disaster, engineering warning devices failed when the rock support failed. Miners may not have the education and experience to assess complicated technical information. Often, computer equipment is not practicable in dusty, dark, and damp underground mines. When local experience can be confirmed by scientific practice, workers have portable, affordable testing methods for assessing hazards. Confirmed by scientific experience, these simple indices provide legitimate warrants for local risk decisions.

The rhetorical transformation of experience is not mere rhetoric.

Since the earliest Roman mines, miners have developed complex vocabularies to describe events and conditions underground (Hoover and Hoover). These vocabularies articulate the spatial and temporal experience of individuals within highly structured institutions that situate individuals differently in relation to risk. Miners, for example, locate themselves in relation to physical objects, events, and conditions. Management views events from a position above and outside the underground world of miners. Miners describe themselves in relation to the "bleeder line" (Willis). Managers call the section "the number one entry of the first left section." The "bleeder" describes a critical component of the mine's ventilation system underground. The "first left section" describes the section in relation to the mine's plan of development, viewed from a systems perspective. The number of the section tells us little about the potential hazard or benefit of a particular location in a crisis. In the 1984 Wilberg Mine fire, for example, the sole surviving miner used the direction of air flow to help him locate an exit (Huntley et al.).

U.S. agencies value local experience and use that experience to control and regulate hazardous worksites, but managers need consistent, quantifiable data to manage risk in large systems. Visual inspection can help miners predict risk in local sites, but local knowledge is sometimes so site-specific that even experienced miners cannot transfer knowledge from one site to another in the same mine (Mark and DeMarco). To predict the consequences of local decisions within large systems, workers must develop the rhetorical flexibility to reconcile local and formal discourses of risk, integrate local and systems perspectives, and develop consistent indices they can apply to assess risk across multiple sites.

The linguistic complexity of multicultural workplaces complicates the processes of rhetorical transformation. Familiar terms in local sites (L1) may not have equivalent terms in formal (L2) risk discourses and vice versa. As table 7.2 suggests, ordinary L1–L2 translations provide an incomplete picture of the four-stage process by which local knowledge and embodied viewpoints (L1) are transformed into the technical language and systems perspective of an L2 discourse of risk (Sauer, forthcoming).

Because spatial relationships are critical in defining and locating hazards, the process of rhetorical transformation also involves what Levelt calls perspective taking—the process of reconstructing visual, kinesthetic, and auditory experience in language so that audiences can apply general knowledge in local sites. This process is necessarily multimodal (see Gentner and Goldin-Meadow; McNeill and Duncan; Sauer *Rhetoric*).

The notion of transformation as an ongoing dynamic, multimodal process can help technical communicators understand how language interac-

	local terms		discourse of risk
L1	L1	⇐norms of transformation⇒	No L1 equivalent
L2	No L2 equivalent		L2

TABLE 7.2

The Process of Transformation

tions and documentation practices may affect the outcomes of technical practices even when those interactions and practices are invisible to participants in the interaction (Gumperz). When we understand the limitations of language as a problem of transcription and documentation (Olson), technical communicators can work with graphic designers, new media specialists, and online information designers to develop new methods for capturing information critical to the regulatory process—information that may exist outside of conventional (agency) documents and procedures. By locating the sites where knowledge is transformed and refigured, cultural analysis allows us to help technical communicators reexamine communication practices critical to risk management and assessment within large regulatory agencies.

The cycle of documentation within institutions is not a closed loop.

Within the cycle of technical documentation in large regulatory industries, cultural history is continually rewritten and reconstructed within agencies and institutions that regulate risk. Old standards must be expunged to protect the health and safety of workers. This need to expunge may be in tension with the need to maintain an archival history of risk regulation that can help regulators build on the past to prevent disasters in the future.

Figure 7.2 does not show, of course, those moments in the cycle when political and economic pressures disrupt the cycle—when administrations deregulate critical industries, for example, or fail to enforce existing regulations. More important, it does not fully show how institutional processes intersect with popular culture and local tradition. Mining—at least in the United States—is a highly regulated industry, even in small mines, which are nevertheless subject to mining regulation. Analysis of the political and institutional dimensions of risk communication can help prepare technical communicators for the negotiations and compromises they may encounter in real-world writing situations.

Social systems resist regulation and inadvertently construct a potentially dangerous archival record in the contexts of risk.

Standards and guidelines live very different lives outside of regulatory institutions that work to control meaning and interpretations critical to safety. The problems of living documents became apparent when I began to check sources and obtain permission for figures and illustrations for my recent study of *The Rhetoric of Risk*. APA citation practices had undergone two revisions in the previous two years to accommodate online source materials. At first, APA required only the URL. In the most recent revision, APA now requires the date of citation. In some cases, when I checked the "same" regulation, the wording had changed online. I cited the 1990 Act in the footnotes, but APA citation practices did not have a format for citing the specific date of revisions to the 1990. Beason's comments are embedded in Bernard's testimony. Beason's original memo appears in the hearings with Bernard's comments and an attached cover letter. It was not surprising that my copyeditor was confused, and APA conventions seemed inadequate for the realities of real-time communication practices.

The problem of living documents affects all aspects of publication practice. When standards change, old standards must be expunged from the regulation; they are unavailable in any "living" (current) copy of the regulations. Yet management is held liable for standards in place at the time of an accident. Old regulations must be available for lawyers and historians, but carefully guarded in practice.

Technical communicators face similar problems in research. They must also find ways to update regulations and standards in online and hard-copy formats, and they must ensure that old regulations and instructions are expunged from the record. Because they are easier to update, online documentation systems seem to offer advantages over hardcopy systems; XTML and HTML systems can create "living documents" that respond to users' needs, reflect changing standards, and even sense changes in the surrounding data environment. But the ease of creating updates may also expunge valuable archival information from the record. The cultural historian can work with technical information designers to sort through the kinds of information that agencies need to understand documentation practices and their effects on policies and procedures. The technical communicator can work with information designers to ensure that documents are accessible, navigable, and available to audiences who need particular kinds of information for particular purposes.

As online capabilities expand, information systems designers are developing new standards for XTML documents that can encode formulas and data processes. These new documents can help readers determine how data was developed, where it came from, and what variables and standards were applied in the process. Such information will be particularly valuable if we are to create transparency in reporting for stocks and other financial transactions. At the same time, however, technical communicators must work with designers to balance the tension between security and safety (on the one hand) and transparency and openness on the other. Cultural theorists must help technical communicators determine how cultural practices also influence how readers create, interpret, and apply data in local contexts.

Postscript

In previous work, I have used (1) feminist analysis to demonstrate how agency documentation practices have silenced the voices of women and Others whose knowledge might contribute to the agency's understanding of disasters (Sauer "Sexual"); and (2) psychological studies of gesture to document how miners use both speech and gesture to integrate scientific theory and embodied practice (Sauer "Embodied Knowledge"). In my most recent work, I am using methods of linguistic anthropology to understand the complexities of workplace discourse in very difficult cross-cultural contexts like South Africa (Sauer "Language"). These studies use cultural theory to create a new understanding of workplace practice.

In the interdisciplinary practice that I imagine, cultural theorists will not merely critique the obvious social and institutional structures that affect the health and safety of workers. Instead, researchers will use methods of cultural analysis to understand why material sites are so difficult to regulate, how documentation functions in these environments, and how specific documentation practices contribute to problems in local contexts. This work extends the work of Geertz, Gumpertz, Haviland, and others who seek to understand how culture is created through talk-in-interaction.

In my current work investigating language practices in South Africa, postcolonialist theory, race, and gender provide powerful lenses for examining workplace practices. But neither race nor gender is sufficient. As analysis of speech and gesture reveals, small differences in language practices, interpreted in the light of past antagonisms, produce misunderstanding and miscommunication that can escalate into violence. If we cannot undo past practices, technical writers can still work to improve communication in complex material and social environments. This inquiry will require thick description and careful analysis of language in contexts of use. There is still much work to be done. As Collins and Olson suggest, written documentation provides an uncertain basis for judgments about risk because writers lack documentation practices to capture the spatial and temporal complexity of real-time workplaces. As the present analysis suggests, written documentation (or its archive) may also provide an inadequate basis for judgments about cultural practice.

Both language and technical environments change over time. Documentation may freeze that moment for analysis, rendering an immediately outdated "snapshot" of reality. The complexity of writing accident narratives demonstrates the difficulties of unraveling the events and conditions that precipitate a disaster, yet workers depend on written instructions and procedures to safeguard themselves and others. Technical documentation is never adequate or complete, yet technical writers must continually work to produce a "more adequate" picture of reality. No single situated viewpoint provides a complete framework for understanding reality.

Technical communicators work in material and institutional sites where relativism may be costly; where there are physical constraints on the production and dissemination of data—regardless of media; and where regulators work to control risks in dynamic, explosive, and highly uncertain physical environments. If cultural theorists critique the limitations of regulations that seem to deny the uncertainty of dynamic and hazardous worksites, technical communicators must also create documents that facilitate decision-making that prevents the (real) deaths of (real) subjects when danger is imminent—without creating additional risk if new genres do not convey meanings embodied in conventional practice.

In an essay examining cultural studies of science as a philosophical program, Rouse argues that cultural studies can help philosophers eliminate the false dualism between the world of practice and the "view from nowhere" (442). Rouse asserts that cultural theorists can bridge the seeming incommensurability of nature and the world of representations if they "insist that meaning emerges within practices, normatively configured interactions with material surroundings" (443).

Rouse argues that science studies are "'internal' to the culture of science" to the extent that "science studies do not come in from the 'outside' to settle the differences among competing *practices*" (452). Instead, there is a "continuity between action and the explicit interpretation and justification of action: action takes place within the presumptive space of reason giving and sense making, and to act in a particular way is also implicitly to construe one's situation in a way that would make sense of one's action within it" (452). "Considerations of power might become relevant to conceptions of language and meaning" without reifying either power or reducing the rhetorical force of language if we understand power dynamically situated in materially circumscribed contexts that enable or constrain action (Rouse 450). To be reflexive, Rouse concludes, critical theorists understand their own practices metadiscursively and honestly: "What do these writings and sayings do? To whom do they speak? What other voices and concerns do they acknowledge, make room for, or foreclose? Which tendencies and alignments do they reinforce and which do they challenge? Above all, to whom are they accountable?" (453).

The question of accountability has particular resonance in the context of the preceding discussion of regulation and risk. When writers construct written standards and regulations, they attempt to articulate the difficult-to-articulate norms that govern risk communication in practice. As we have seen above, archival histories often do not document how norms evolve and change over time or how outdated norms and procedures persist in texts that circulate outside of the institutions that control meaning in practice. In each case described above, risk knowledge is continually reconstructed in the agreements that take place among individuals in an ongoing cycle of review and revision within large regulatory agencies.

Ultimately, researchers in technical communication and cultural studies can learn much from each other. To the extent that regulations support that work, technical communicators must be particularly careful in the assertions they make about culture, power, and the representational character of language. They must look reflexively and critically at the evidence they draw on to warrant assertions about language; they must develop methods to identify gaps in the record, and they must understand how the available archival records may skew the narratives they construct from necessarily incomplete archival histories.

Notes

1. As I demonstrate, for example, coal mine accident investigations may invisibly encode assumptions about the status of individuals within institutions and environments that locate individuals differently in relation to risk ("Sexual").

2. This research was funded by generous grants from the National Science Foundation's Program in Ethics and Values Studies in Science and Technology. I am especially grateful to Rachelle Hollander, Program Director, for her encouragement in this project.

3. Schnakenberg, G. (Personal interview, 2001) first suggested the notion of "living documents."

4. Federal Mine Safety and Health Act of 1977, 30 CFR § 75(202), p. 500. I refer to the July 1, 1990, version of standards in this passage. Because the Mine Act is a living document, it is possible to speak of specific sections without identifying the year in which these provisions were created. Thus, the 1969 Act and its amendment

will always be the current set of regulations in force at any one time. If regulations change, the Act is updated with the new standard or standards. The old standards no longer exist in the public record because outdated standards could endanger the health and safety of workers. The research for this project is based on the standards available in 1990, except when I cite a specific iteration of the Act. The reader should be warned that 1990 standards may no longer be in force in the current iteration of the Act. It is also technically incorrect to speak of a 1990 Act. Citations in the text are intended to help orient the reader to the specific wording that might have caused confusion at specific moments in time. The citations are not intended as legal citations to authority.

5. Bernard argued that "Protection is provided on these rare occasions by the requirements of Sections 103 j) and k) of the Act which assures appropriate control in emergencies and by the mine standards of Part 49" (490).

WORKS CITED

Bazerman, Charles. *The Languages of Edison's Light.* Cambridge, MA: MIT P, 1999.

Bazerman, Charles, and David Russell. *Writing Selves/Writing Societies: Research from Activity Perspectives.* Fort Collins, CO: WAC Clearinghouse and Mind, Culture, and Activity, 2002. http://wac.colostate.edu/books/selves_societies/ (1 February 2003). Accessed 4 July 2004.

Beason, R. L. Memorandum for: Alan C. McMillan, Acting Assistant Secretary, Arlington, VA, Re: Gassy Mine Standard Changes dated 12 June 1986. U.S. House of Representatives, 100th Congress 1st Session, Committee on Labor and Human Resources, Oversight of the Mine Safety and Health Administration. Hearings before the Committee on Labor and Resources . . . on examining activities of the Mine Safety and Health Administration (pp. 478–491). Washington, DC: USGPO, 1987: 475–491.

Beck U., Bonss, W., and Lau, C. "The Theory of Reflexive Modernization: Problematic, Hypotheses and Research Programme." *Theory, Culture and Society* 20.2 (2003): 1–17.

Berkenkotter, C. and Huckin, T. *Genre Knowledge in Disciplinary Communication: Cognition/Culture/Power.* Hillsdale, NJ: Lawrence Erlbaum, 1995.

Bernard, R. L. Memorandum for: Ronald L. Beason, Metal and Nonmetal Mine Safety and Health Inspector Re: Changes to Gassy Mine Standards dated 12 June 1986. U.S. Congress House Committee on Labor and Human Resources. Oversight hearings on the Mine Safety and Health Administration. 100th Congress, 1st sess. Washington, DC: GPO, 187. 478–491.

Brown, R. H. "Modern Science and Its Critics: Toward a Post-positivist Legitimization of Science." *New Literary History* 29.3 (1998): 521–550.

Chlorine Chemistry Council (1996/2003). Questions and answers (FAQ: Dioxinfacts.org). [Online] Available: http://www.dioxinfacts.org/questions_answers (Accessed 30 December 2003).

Clarke, L., and Short, J. F. "Social Organization and Risk: Some Current Controversies." *Annual Review of Sociology* 19 (1993): 375–399.

Code, L. *What Can She Know? Feminist Theory and the Construction of Knowledge.* Ithaca and London: Cornell UP, 1991.

Collins, H. "The Structure of Knowledge." *Social Research* 60 (1993): 95–116.

Curran, D. J. *Dead Laws for Dead Men: The Politics of Federal Coal Mine Health and Safety Legislation.* Pittsburgh: U of Pittsburgh P, 1993.

Dawes, R. M. "Irrationality in Everyday Life: Professional Arrogance and Outright Lunacy." Unpublished manuscript, 2001

———. *Rational Choice in an Uncertain World.* New York: Harcourt Brace, 1988.

Dombroski, Paul. "The Lessons of the Challenger Investigation." *IEEE Transactions on Professional Communication* 34.4 (1991): 211–216.

Dubow, S. *Racial Segregation and the Origins of Apartheid in South Africa, 1919–36.* New York: St. Martin's Press, 1989.

Engstrom, Y. *Understanding Practice: Perspectives on Activity and Context.* Cambridge: Cambridge UP, 1996.

Geertz, Clifford. *The Interpretation of Cultures: Selected Essays.* New York: Basic Books, 1973.

Gentner, D. and Goldin-Meadow, S., eds. *Language in Mind: Advances in the Study of Language and Thought.* Cambridge, MA: MIT P, 2003.

Gumperz, J. J. "Interviewing in Intercultural Situations." *Talk at Work: Interaction in Institutional Settings.* Ed. P. Drew and J. Heritage. Cambridge: Cambridge UP, 1992. 302–327.

Haviland, J. B. "Projections, Transpositions, and Relativity." *Rethinking Linguistic Relativity.* Ed. J. Gumperz and S. Levinson. Cambridge: Cambridge UP, 1996. 271–323.

Herndl, C. G., Fennell, B.A., and Miller, C. "Understanding Failures in Organizational Discourse: The Accident at Three Mile Island and the Shuttle Challenger Disaster." *Textual Dynamics of the Professions.* Ed. Charles Bazerman and James Paradis. Madison: U of Wisconsin P, 1991.

Hoover, H. C. and Hoover, L.H. Introduction. *De Re Metallica.* G. Agricola. Trans. H. C. Hoover and L. H. Hoover (from the first Latin Edition of 1556; reprint of the 1912 translation published by *The Mining Magazine*, London). New York: Dover, 1950.

Huntley, D. W., Painter, R. J. , Oakes, J. K., Cavanaugh, D. R., and Denning, W. G. Report of investigation: Underground coal mine fire. Wilberg Mine. I.D. No. 42000080. Emery Mining Corporation. Orangeville, Emery County, Utah. 19

December 1984. Arlington, VA: U.S. Dept. of Labor, Mine Safety and Health Administration, 1992.

Irwin, A. *Citizen Science: A Study of People, Expertise, and Sustainable Development.* London: Routledge, 1995.

Irwin, A., and Wynne, B. *Misunderstanding Science? The Public Reconstruction of Science and Technology*. Cambridge: Cambridge UP, 1996.

Knorr, Karin D. "Producing and Reproducing Knowledge: Descriptive or Constructive? Toward a Model of Research Production." *Social Science Information* 16.6 (1977): 669-696.

Latour, Bruno. "Is Re-modernization Occurring—And If So, How to Prove It? A Commentary on Ulrich Beck." *Theory, Culture & Society* 20.2 (2003): 35–48.

———. *We Have Never Been Modern.* Trans. C. Porter. New York: Harverster Wheat Sheaf, 1993.

Latour, Bruno, Steve Woolgar, and Jonas Salk. *Laboratory Life*. Princeton, NJ: Princeton UP, 1986.

Levelt, W. J. M. "Perspective Taking and Ellipsis in Spatial Descriptions." *Language and Space*. Ed. P. Bloom, M. A. Peterson, and M. F. Garrett. Cambridge, MA: MIT P, 1996. 76–107.

Longo, Bernadette. *Spurious Coin: A History of Science, Management, and Technical Writing*. Albany: SUNY P, 2000.

Lynch, M. *Art and Artifact in Laboratory Science: A Study of Shop Work and Shop Talk in a Research Laboratory*. London: Routledge, 1985.

Lynch, M. and Woolgar, S., eds. *Representation in Scientific Practice*. Cambridge, MA: MIT P, 1990.

Mark, C., and DeMarco, M. J. "Longwalling Under Difficult Conditions in U.S. Coal Mines." *CIM Bulletin* April (1993): 31–38.

McAteer, J. D., and Galloway, L. T. Testimony...on behalf of the Council of Southern Mountains, Inc. (pp. 554–715). In U.S. Congress House. Subcommittee on Labor Standards of the Committee on Education and Labor. Oversight hearings on the Coal Mine Health and Safety Act of 1969 (Excluding Title IV). 95th Cong., 1st sess. Washington, DC: GPO. 29 June 1977.

McCloskey, D. *The Rhetoric of Economics*. 2nd ed. Madison: U of Wisconsin P, 1998.

McNeill, D. and Duncan, S. D. "Growth Points in Thinking-for-Speaking." *Language and Gesture: Window into Thought and Action*. Ed. D. McNeill. Cambridge: Cambridge UP, 2000. 141–161.

Molinda, G., and Mark, C. Coal mine roof rating (CMRR): A practical rock mass classification for coal mines. U.S.MB IC 9387 (Bureau of Mines information circular/1994 C 9387). Washington, DC: U.S. Department of the Interior, Bureau of Mines, 1994.

Moodie, T. D. and Ndatshe, V. *Men, Mines, and Migration*. Berkeley: U of California P, 1994.

Noah, Lars. "Divining Regulatory Intent: The Place for a 'Legislative History' of Agency Rules." Abstract. *Hastings Law Journal* 51.2 (2000): 255+. Available: http://papers.ssrn.com/sol3/papers.cfm?abstract_id=212748#PaperDownload.

Olson, D. R. *The World on Paper*. Cambridge: Cambridge UP, 1994.

Perrow, C. *Normal Accidents*. New York: Basic Books, 1984.

Petroski, H. *To Engineer Is Human: The Role of Failure in Successful Design*. New York: St. Martin's P, 1985.

Rouse, Joseph. "Understanding Scientific Practices: Cultural Studies of Science as a Philosophical Program." *The Science Studies Reader*. Ed. M. Biagioli. New York: Routledge, 1999. 443–456.

Sauer, Beverly. "Embodied Experience: Representing Risk in Speech and Gesture." *Discourse Studies* 1.3 (1999): 321–354.

———. "Embodied Knowledge: The Textual Representation of Embodied Sensory Information in a Dynamic and Uncertain Material Environment." *Written Communication* 15.2 (1998): 131–169.

———. "Language Acquisition and the Discourses of Risk." *Chicago Linguistics Society Proceedings* (Proceedings of the 39th Conference of the Chicago Linguistics Society, Chicago: 7–10 April 2003). Forthcoming.

———. *The Rhetoric of Risk: Technical Documentation in Hazardous Environments*. Mahwah, NJ: Lawrence Erlbaum, 2003.

———. "Sexual Dynamics of the Professions: Articulating the Ecriture Masculine of Science and Technology." *Technical Communication Quarterly* 3.3 (1994): 309–12.

Schnakenberg, G. Personal interview. Pittsburgh, PA. April 2001.

Selzer, Jack, ed. *Understanding Scientific Prose*. Madison: U of Wisconsin P, 1993.

Trento, J. *Prescription for Disaster*. New York: Crown, 1987.

U.S. Dept. of Labor, Mine Safety and Health Administration. Federal Mine Safety and Health Act of 1977, Public Law 91-173, as amended by Public Law 95-164. [Online] Available: http://www.msha.gov/regs/act/acttc.htm Accessed 19 December 2003. U.S. Presidential/Congressional Commission on Risk Assessment and Risk Management. (1997). Framework for Environmental Health Risk Management—final report (Vol. 1). Washington, DC: USGPO, 2003.

Vaughan, Diane. *The Challenger Launch Decision: Risky Technology, Culture, and Deviance at NASA*. Chicago: U of Chicago P, 1996.

———. "The Dark Side of Organizations: Mistake, Misconduct, and Disaster." *Annual Review of Sociology* 25 (1999): 271–305.

———. "Rational Choice, Situated Action, and the Social Control of Organizations." *Law & Society Review* 32.1 (1998): 23–61.

———. "The Trickle-down Effect: Policy Decisions, Risky Work, and the Challenger Tragedy." *California Management Review* 39.2 (1997): 80–102.

Walzer, Arthur E., and Alan Gross. "Positivists, Postmodernists, and the Challenger Disaster." *College English* 56.4 (1994): 420–433.

Willis, I. J. (1993). Investigation. In Re: Southmountain Coal Company, Inc. No. 3 Mine. Wise County, Norton, VA. Explosion of Dec. 7, 1992 (Deposition). Southmountain, VA: U.S. Mine Safety and Health Administration/ Virginia Department of Mines.

Winsor, Dorothy. "The Construction of Knowledge in Organizations: Asking the Right Questions about the Challenger." *Journal of Business and Technical Communication* 4.2 (1990): 7–20.

Pedagogy

As practice is never theory-free, teachers consider the theory-to-practice arcs that their pedagogies forward. The four chapters in this section resist any split between (cultural) theory and (technical communication) practice or, similarly, between critical reflection and pragmatic action. Together, they illustrate the usefulness, even the necessity, of cultural theory for empowering our students and those with whom they engage (or will engage) as technical communicators. At the same time, the chapters suggest ways that technical communication can inform and situate the work of cultural studies.

Jim Henry offers an updated version of his article "Writing Workplace Cultures—Technically Speaking," first published in *College Composition and Communication Online*. Henry argues that modernist modes of composition instruction (including process pedagogy) must be replaced with postmodern ones that can account for the changes that globalization, or "fast capitalism," has wrought on the workplace. While many technical communicators, along with other workers, are taking on more hours, more responsibilities, and more complex tasks—all in response to a call for more "flexible" labor—they are losing job security, control over the terms and conditions of their employment, and professional status. Further, these new conditions and roles make it extremely difficult for technical communicators to mobilize on their own behalf.

It follows that we need new forms of resistance and, to develop these, new theoretical frameworks and pedagogical approaches. One such approach that Henry employs is autoethnographic inquiry, which asks students to critically examine how the cultures and discourses where they work position them as technical communicators and to what effects. Henry's students then produce *petits recits*, or "small stories," that critique and rewrite these subjective work identities and thereby function as counternarratives to those of fast capitalism.

Importantly, Henry and his students do not just perform typical studies of organizational culture but also (1) link their studies "of local conditions to global [systemic] social structures" (203) and, (2) seek ways to resist disempowering practices and the identities that they maintain (213). We can foster this resistance, Henry explains, by helping our students develop "an aptitude for cultural criticism, grounded in skills at cultural analysis in real settings," and by helping them form workplace alliances (215).

Like Henry, Michael Salvo, in "Rhetoric as Productive Technology," argues for critical pedagogies that position students to enact ethical change in the workplace. Building on James Berlin's call to help students become "active agents of social and political change," Salvo adds that we must also help them enact technological change: "[t]echnical rhetoricians have the unique capacity to contextualize combinations of technologies and social relations . . ." (226). Additionally, Salvo refines Berlin's notion of technoculture as stated in *Rhetorics, Poetics, and Cultures*, while nodding to the contributions of Feenberg, Barry, and Ulmer. For Salvo, like Henry, students' critique of the interrelations among humans, technologies, and institutions must not paralyze them in dismay but mobilize them to design alternative interrelations. This mobilization, which involves working with (not for) communities to create meaningful, usable, intelligent solutions, is the work of what Salvo calls the "postmodern expert." Salvo's own research and pedagogy on user-centered engagement illustrates the postmodern expert's roles, but here he elaborates through a detailed review of recent studies by Brenton Faber, Jim Henry, Beverly Sauer, Teresa Kynell-Hunt and Gerald Savage, and David Kaufer and Brian Butler.

In arguing for the productive potential of cultural critique, Salvo is also careful to explain how cultural studies can take cues from technical communication. "Focused as it is on the workplace," he writes, "technical and professional communication is uniquely able to . . . inform postindustrial labor policy as it understands the discourses of workplace production, of interrelations of individuals, values, ideologies, and the discourses they produce" (223).

In his chapter on cultural studies' contributions to technical communication service-learning pedagogy, Blake Scott similarly argues for the movement from cultural critique to ethical civic action but also shows how this movement can be short-circuited by hyperpragmatist forces. After explaining how this short-circuiting can occur, especially in the "reflection" component of service-learning, Scott turns to cultural studies as a possible anecdote. Cultural studies models, concepts, and concerns, Scott argues, can help students account for, critique, and ethically revise the broader cultural conditions, circulation, functions, and effects of their service-learning work.

Scott's service-learning pedagogy draws on such cultural studies models as the cultural circuit and intercultural inquiry to reconfigure reflection and invention activities, usability strategies, and major assignments. Most notably, Scott transforms a final report assignment into an "action plan" in which students recommend ways to "ethically implement, revise, or build on" the documents and relationships created through their projects (253). Scott's students gain the practical experience of negotiating complex, real-world rhetorical situations, but they also gain the opportunity to model Salvo's notion of the "postmodern expert," ethically engaged with community partners in an effort to create more empowering and responsive processes and texts. Scott not only shows how cultural studies can ensure service-learning's ethical potential, he also shows how service-learning approaches to technical communication can provide students with contexts for implementing, and not just generating, cultural critique.

Finally, Katherine Wills, in "Designing Students," powerfully amplifies the other chapters' calls for cultural studies approaches that resist both hyperpragmatism, driven partly by anti-intellectualism, and critical passivity. Wills also points to another "workplace" context in which technical writing students and the graduate students who train them can critique and respond to hegemonic power: the classroom. Drawing on Bourdieu's *Homo Academicus* to explain how "pseudotransactional" modes of instruction are rooted in and maintain academic cultural capital, Wills exhorts us to apply alternative pedagogies based on the sociopolitical analysis of power. Like Elizabeth Britt does earlier in the collection, Wills investigates the influence of power in "hegemonic systems and rule-governed contexts" (259). Britt and Wills similarly argue that neither research nor pedagogy is immune to the "accumulation of . . . capital that reifies underlying extant hierarchies" (260)—whether in Bourdieu's academy or Schryer's insurance company. Alternative modes of instruction, Wills adds, must be responsive to the "highly interconnected, global, and multicultural paradigms" of our day (260).

Wills offers readers, instructors, and graduate teaching assistants both pedagogical and conceptual strategies for revising pseudotransactionalist instruction with cultural studies. These strategies include foregrounding how power relationships and ideological agendas operate in a variety of cultural frames (including international ones), presenting course syllabi and curricula as sites for critiquing academic power hierarchies, and creating assignments that require abstraction, systems thinking, and experimentation. Wills echoes Greg Wilson's (and Henry's) assertion that the sociopolitical analysis of power will prepare students to be better and more effective corporate citizens.

Chapter Eight

Writing Workplace Cultures—Technically Speaking

Jim Henry

KALI: Here's the setting: battleship-gray walls with rose flecks that are an attempt to add cheer but end up looking like bloodstains. The floor plan is obviously cubes, but they're attached in such a way as to resemble a maze as well. Everyone is walking around holding clipboards. Alan enters in a whirl, like a Tasmanian Devil.

ALAN: Has anyone seen Mary? I need to find Mary. I need to *fire* Mary. I don't want to, but someone upstairs messed up and I have to. Tell Mary I'm looking for her.
(He exits; Mary enters.)

MARY: Has anyone seen Alan? He needs to fire me so I can get back to work. I need to get all this done before I go.
(Alan reenters.)

ALAN: Mary. There you are. You're fired.

MARY: I know.

ALAN: I don't want to be the one to tell you but I have to.

MARY: I know.

ALAN: It's nothing you've done.

MARY: I know.

ALAN: You're a great employee.

MARY: I know.

ALAN: Three weeks.

MARY: I know.

Another version of this chapter originally appeared in *College Composition and Communication Online* 53.2 (2001). Reprinted with permission of the National Council of Teachers of English

In a workshop devoted to fictional performative ethnographies, a practicing technical writer composed this one-act, one-scene play depicting her workplace. The workshop was part of Cultures of Professional Writing, a course I designed in George Mason University's M.A. Program in Professional Writing and in which participants conducted ethnographic work aimed at analyzing literacy practices in local writing cultures. "Kali" is the pseudonym chosen by this participant as she collaborated in "informed intersubjective research" (McBeth), serving as my research subject along with other course participants as they researched organizational and discursive features of workplace cultures. Their work now figures as part of an archive available at the following URL: http://archive.ncte.org/ccc/2/53.2/henry/index.html.

Kali's play was only a brief foray into the realm of fiction in a course otherwise devoted to nonfiction writing grounded in social science methodology, yet this play on work elicited nods of acknowledgment among her peers. For good reason: In the eleven years that I have taught this course, at least one participant has either been fired or has quit her job during the semester. I interpret this fact as an indicator of the ways in which "flexibility of labor," one of the characteristics of U.S. economy in the era of globalization, augurs poorly for professional writers. In fact, "flexibility" probably augurs poorly, in many ways, for the majority of Americans who will enter the workforce in the years to come. That my example comes from the professional class of writers, those workers who daily enact composition tenets as their central workplace duty, should give all of us in the fields of composition and technical communication pause. Our students enter a work world in which discursive processes and products have been radically transformed by macroeconomic and technological shifts of the last decade, shifts with which our composition and technical writing epistemologies, curriculum design, and course conceptualization have yet to contend.

COMPOSITION'S SUBJECTS

Histories of composition during the twentieth century inevitably trace the shift in emphasis from product to process in research and in the classroom. First came a focus on cognitive processes during the 1980s, which gave way to a focus on social processes in the eighties and early nineties. The purported "paradigm shift" was welcomed almost universally, it seemed, so exhausted was the intellectual potential in assigning demonstration essays, so pointless the evaluating of writing only after complete, and so frustrating the conceptualizing of writing as something as simple as communicative packaging. Reasons why composition had for so long focused on product have been amply supplied by histories of our field that trace curricular developments at influential colleges and universities which then set models to be followed (see, e.g., Berlin, *Rhetorics*; Connors; Russell). Less investigated have been the correlations between composition epistemology and workplace literacy demands as shaped by the twentieth-century U.S. economy and its modes of production.

A short version of these correlations goes something like this: in the early twentieth century emerged the first U.S. economies of scale, by which industrial innovations such as the assembly line enabled mass production (and consumption) that witnessed the simultaneous demise of much guild and craft production. To improve "productivity," principles of "scientific management" as espoused by Frederick W. Taylor were enacted. Using time-motion studies, workers' movements were standardized, work processes were routinized, and corporate profits hence maximized by the elimination of endeavors that did not apparently contribute directly and unequivocally to the product. Literacy tasks were similarly truncated, as explained by JoAnne Yates:

> Systematic management as it evolved in the late nineteenth and early twentieth centuries was built on an infrastructure of formal communication flows: impersonal policies, procedures, processes, and orders flowed down the hierarchy; information to serve as the basis for analysis and evaluation flowed up the hierarchy; and documentation to coordinate processes crossed the hierarchy. These flows of documents were primary mechanisms of managerial control. (20)

Hence, the writing demanded of most workers was a product that contributed to other products, as managed in the workplace. Writing processes (and, in retrospect, the worldly concerns of the writing subject) were not only irrelevant but effectively off limits according to the new scientifically managed workplace cultures.

David Russell discusses curricular innovations as correlated with emerging cultures of professionalism during the same period, observing that "writing was the oil that kept the new industrial machine running smoothly" (104). He traces turf battles and struggles within colleges and universities for dominion over writing, and he notes the emergence of the "term paper" in the 1930s simultaneously with specialization in the university not unlike the specialization that Taylorization had wrought on shop floors. (Full-time professors were now judged less by their teaching and mentoring and more by their publications.) The research paper, for example, previously a mechanism for initiating students into their future intellectual and professional realms, was now often assigned by instructors who knew little of those realms. As Russell puts it, "As the practice of assigning research papers settled into routine, its pedagogical uses seem to have narrowed, moving from apprenticeship to production" (89). This emphasis on "production" set the scene for composition's emphasis on "product," a scene that was equally sustained by hiring and staffing policies at universities that endure to this day. Teaching in a stand-alone classroom with one instructor per *X* number of students, emphasizing singly authored "essays" "submitted" for a grade (which were, and are, most often normed), and ignoring such topics as the management of the classroom, the relationships among students, the purview of instructors, and the investments of the university, effectively assured a writer-worker with appropriate literacy practices for the lower and middle echelons of the twentieth century "high-volume" mode of production.

The shift in composition to a focus on process resulted at least in part from the dismay of writers at the very conceptualization of writing inherent in the product approach. Why wouldn't we want to afford student-writers the "real" experience of writing as we knew it—as an encounter with readers, as a response to feedback, as a route to new visions? At stake in part was the vital issue of writerly sensibilities, what we knew them to be, and our conviction that to teach composition must surely entail conveying something about these sensibilities. Hence, the shift to process that in theory would enable such conveyance. By the late 1980s, process approaches had gone beyond the initial focus on cognitive processes to the social, and with this evolution composition's terrain became fraught with issues of identity as mediated through language. During the same years, the rosy picture of process pedagogy clouded, due in large part to the administrative practices at most colleges and universities that positioned Composition as an endeavor focused not on writerly sensibilities but on some new version of product, enacting a pedagogy often referred to in the literature as current-traditionalism. Sharon Crowley comments:

> The easy accommodation of process-oriented strategies to current-traditionalism suggests that process and product have more in common than is generally acknowledged in professional literature about composition, where the habit of contrasting them conceals the fact of their epistemological consistency. A truly paradigmatic alternative to current-traditionalism would question the modernism in which it is immersed and the institutional structures by means of which it is administered. Process pedagogy does neither. (212)

Crowley's misgivings resonate strongly and help explain the forays into postmodernism undertaken by compositionists in recent years. By way of example, Susan Miller has proposed "disclosing connections between specific social and textual superstructures and highlighting how writing situations construct their participant writers before, during, and after they undertake any piece of writing" (198). Similarly, Lester Faigley has pondered postmodernity and the "subject" of composition, leading him to suggest that students be invited to explore "how agency can be constructed from multiple subject positions" (224). The terminology of postmodernism is rich with references to textual and social structures and practices and with a relentless focus on the "subject," a theoretical entity posited in order to depict lived experience as shaped by language practices and by ideological forces. This subjectivity shares much terrain with writerly sensibility, that elusive phenomenon sought through process pedagogy yet only ever partially perceived under that paradigm because of its failure to go beyond modernist epistemology and to break out of conventional university structures.

Key to postmodern inquiry into subjectivity is the notion of *discourse*. Modernist understandings of discourse see it as a vehicle, as words chosen to convey preexisting ideas and thoughts and which might be said to mark a speaker or speakers as belonging to a specific social group, as in the case of a "discourse community." Postmodern versions of discourse see language as the very material from which reality is formed: we are born into language and we learn notions of self (and other) only by way of the many discourses we encounter and which provide us with the means to understand and interpret reality. In life, we occupy sequential and overlapping "subject positions" in discourse that open and restrain avenues of action and thinking, and we are afforded these positions differently and unequally, depending on a host of ideological forces. Only by integrating such theory into composition epistemology and practices, say postmodern compositionists, can we truly begin to enable student writers to perceive the complexity of their writerly sensibilities as they shape discourses and as discourses shape them.

Pondering the writing subject under the aegis of postmodernism is enhanced by other tenets that come with this theoretical terrain: the need to embrace the quickening of cultural processes and the reshaping of subjectivity through technology (Virillo; Selfe and Selfe; Johnson-Eilola); the need to resist the totalizing narratives of modernism, in Jean-François Lyotard's version by seeking out *petits recits* that offer counternarratives; the importance of linking the "small stories" of local conditions to global social structures in specific historical moments, the better to probe class relations (Jameson); the need to inspect a central tenet of the new capitalism, "flexibility of labor," for its implications for those workers outside management circles (Harvey); and perhaps most important, the need to embrace issues of agency as globalization and technology render agency ever more difficult to discern (Bauman). The next section takes up some of these issues as they might prove vital for the field of composition and the related field of technical communication as they shape future workers.

SUBJECTS OF THE NEW WORK ORDER

The nature of the workplace has changed dramatically in the fifteen years since the contributors to Lee Odell and Dixie Goswami's *Writing in Nonacademic Settings* opened doors for Composition. Their studies of textual and social structures in the U.S. workplace in the early 1980s offered glimpses at intricate and complex processes behind corporate products at a time when corporations themselves were transmogrifying. The high-volume economy that had determined organizational structures and literacy requirements earlier in the century was giving way to a "high-value" economy in which the service sector would replace manufacturing at the core of the U.S. economy. Former U.S. Secretary of Labor Robert Reich describes the state of affairs:

> America's core corporation no longer plans and implements the production of a large volume of goods and services; it no longer owns or invests in a vast array of factories, machinery, laboratories, warehouses and other tangible assets; it no longer employs armies of production workers and middle-level managers; it no longer serves as gateway to the American middle class. In fact the core corporation is no longer American. It is, increasingly, a facade behind which teems an array of decentralized groups and subgroups continuously contracting with similarly diffuse working units all over the world. (81)

This trend in world economies has come to be known as "globalization," the deepest meaning of which, claims Zygmunt Bauman, is "that of the indeterminate, unruly and self-propelled character of world affairs; the absence of a centre, of a controlling desk, of a board of directors, of a managerial office" (59). U.S. workplaces, in the era of high-value production in the age of globalization, require workers very different from those of just two decades ago. James Paul Gee, Glynda Hull, and Colin Lankshear trace the ideologies of "new" or "fast" capitalism, noting that the knowledge, information, and responsibility formerly given to managers is now (in theory) pushed down to front-line workers:

> This, however, requires workers now who can learn and adapt quickly, think for themselves, take responsibility, make decisions,

> and communicate what they need to leaders who coach, supply, and inspire them.... Gone then—except, again, in the backwaters of the old capitalism—are workers hired from the neck down and simply told what to do. (19)

Gone as well is job security, along with many of the other features of work under the old capitalism that afforded workers a say in their work lives, their local work cultures, and even national work culture via unions. (Reich notes that union memberships in the nonagricultural sector plummeted from 35 percent to 17 percent from 1960 to 1990 [6].)

Critical in these developments is what Bauman calls the "absence of a centre." In the earlier economic age, that centre would have been interrelated with national and civic concerns in ways that held it accountable to the polity. In the age of globalization, however, this accountability is significantly dissipated if not wholly lost. Bauman quotes Albert J. Dunlap, who, much as Frederick Taylor shaped the high-volume enterprises, has come to be known as the "rationalizer" of the modern enterprise: "The company belongs to people who invest in it—not to its employees, suppliers, nor the locality in which it is situated" (6). With the capacity for instantaneous, long-distance investment and divestment afforded by the World Wide Web, and with the complete lack of geographical, national, or civic ties between the new owners and the new workers, the "subjectivity" of writers in workplace settings has changed irreversibly.

Consider this subjectivity in tandem with the modernist notions of writerly sensibility implicit in twentieth-century composition epistemology. The educated writer, in this epistemology, by virtue of his or her rhetorical savvy and elevated literacy, would be able to draw on his or her sense of moral responsibility and code of ethics (most often grounded in liberal humanism) to assume agency, to effect changes in the community and the workplace, through democratic processes undergirded by the bond between government (at the local, state, and national levels) and the corporation. But globalization has broken this bond. And at the same time, workers in this new scenario find themselves obliged to work more and more hours (under the constraints of "flexibility") and to retrain constantly, limiting their time for civic engagement that earlier composition epistemology presumed. Otherwise stated, the "subject" we imagine under twentieth-century composition epistemology has become an anachronism.

What to do? Writing in 1993, Mary Beth Debs offered an observation that can serve as prompt: "Although we may want to be cautious in recognizing it as such, the corporation, certainly the organization, has become the major arena for public life for the individual in modern Western civilization" (161). In the light of the foregoing discussion, the "public life" of writers in organizational settings appears dramatically predetermined (even as many of those organizations exert increasing influence on public policy via campaign contributions and PACs), presenting us with a paradox: we must educate technical communicators to exercise their craft in such settings, all the while equipping them with knowledge and know-how that enables a broad and deep understanding of the many forces that shape their writerly sensibilities. Gee, Hull, and Lankshear state it succinctly: "In our view, the new work order is largely about trying to create new social identities or new kinds of people" (xiii). The next section looks at how we might investigate the connections among these new identities for technical communicators, writing practices, and writerly sensibilities.

AUTOETHNOGRAPHIC INQUIRY INTO SUBJECTIVE WORK IDENTITY

In their pioneering study of social and textual structures in workplace writing, the researchers in *Writing in Nonacademic Settings* employed ethnographic techniques, an approach that has proven productive in numerous studies of workplace writing since. (See, for example, Dias, Freedman, Medway, and Paré; Duin and Hansen; Fearing and Sparrow; Garay and Bernhardt; Hull; Lay and Karis; Matalene; Spilka; Sullivan and Dautermann.) This approach is not without pitfalls, however, as Carl Herndl pointed out in *College Composition and Communication* in 1993: "[T]his research lends itself to a mode of reporting that reproduces the dominant discourse of its research site and spends relatively little energy analyzing the modes and possibilities for dissent, resistance, and revision" (349).

When I first offered the course Cultures of Professional Writing in 1993, I was conscious of these pitfalls and eager to support students in imagining revisions of the dominant discourses of worksites, when necessary. I was also intrigued with positioning students as researchers in workplace settings, so that they might learn far more through the process than they could ever learn from reading other researchers' findings. In the first course offering, I anticipated that students would research worksites that were foreign to them, but the on-site time required by fieldwork combined with the desire among student-practitioners to study their own places of work (nearly all of them were employed full-time as professional writers or editors while completing their M.A.'s) led most to conduct autoethnographies. I was only slightly familiar with this subgenre, and since we had adopted in our course an "informed intersubjective" methodology that beckoned me to build upon students' leads, I have examined autoethnography's history, uses, and potential.

I have taught this course every spring since 1993, collaborating with students in their research and using our findings to compose my book *Writing Workplace Cultures: An Archaeology of Professional Writing* in 2000. My subtitle reflects the conceptualization of the book as a resource that others might use in the ways we visit archaeological sites: by comparing artifacts and their relationships in one site to other sites with which one is familiar, one can spark further insight on topics and themes under consideration. The companion website to this book is still available to those who would like to conduct such hypertextual comparisons at the following URL: http://archive.ncte.org/ccc/2/53.2/henry/index.html.

On this site you will find a brief description of its rationale (under the Legend), a master file of abstracts, listed alphabetically by students' chosen pseudonyms, and various tables making connections among their research and their workplace writing experiences. My hope is that you can extend and augment my observations below by perusing the site and making connections to sites closer to home.

I have used the heading "subjective work identity" as an allusion to an insight mustered by Françoise Lionnet, who has used the term "autoethnography" to describe Zora Neale Hurston's autobiography *Dust Tracks on the Road.* By Lionnet's accounting, this autoethnography defines Hurston's "subjective ethnicity" (383). That is, the category of ethnicity, defined on the basis of shared characteristics among a people, was inflected by the ways in which this particular "subject" positioned herself with respect to that ethnicity and its representations. In the Cultures of Professional Writing project, I came to think of my students' autoethnographic writings as having to do with their own "subjective work identity." Their analyses of workplace cultures have given them the opportunity to see how features of their work identities are shared with the other people in the room. Through their work, we have been able to discern shared characteristics of the professional class of writer, which Bureau of Labor statistics cast as comprised in the majority by women (60%), 90–97 percent white (depending on which subcategory of writer), and by 2006, 350,000 strong (Table 2 and Table 11). Technical communicators constitute 50,000 of these workers, and by 2012, the Bureau of Labor estimates that they will be 63,000 strong (21). Some of the shared characteristics as a professional class, based on our findings through autoethnography, are as follows:

Second Class Status in the Organization

Recurrently, these writers characterized their status as "second class" (Ana, Bernie, and Rachel explored this issue at some length) or as "support staff" (as documented by Carolyn, Ella, Ginger, Jeff, and Liz). In May's case, she became aware through her research that her effective status within the organization came nowhere near her status as represented on the org chart, at least compared with the organizational purview of other employees at the same level. We have discussed reasons for this status over the years, and most often students agree that it derives from the commonly held belief that writers contribute little or nothing to the "value added" to organizational products or services. Technical writing, in this view, is merely communicative packaging, and those who conduct it do not merit first-class status.

When students bring the organizational charts of their workplaces (a required assignment), the boxes containing tech writers are invariably at the bottom. And as these autoethnographers all know firsthand, communication circuits among the boxes at the top frequently leave them out of the loop. These cultures evoke the image of "We want robots!" cited by Slack, Doak, and Miller as they recount a corporate recruiter's comments while seeking tech writers. And whereas most of us in the academy would now agree that the technical communicator is indeed an "author," in the Foucaultian sense, most workplace cultures clearly would not. Work remains to be done to correct this underestimation of technical writing's potential and prowess.

The Role of Talking Handbook

Adria, Joe, and Ravi all noted the representations of themselves as "talking handbooks" among their colleagues. That is, when approached by these colleagues, they were invariably plied for a grammatical ruling or syntactical arbitration. Issues of organizational goals or developmental projects were reserved for talks with workers in other positions and with different job titles. In class discussions, other writers concurred, and we reasoned that such a representation surely derived at least partly from a formalist epistemology that their colleagues would have encountered in Composition 101 and in mass cultural representations. When one's expertise is reduced to the formal, one's importance in the organizational endeavor can be nothing more than marginal, and so part of our resolve in class discussions has been to engage with these representations and reveal their inadequacy.

In this sense, our autoethnographic work evokes the kind described by Mary Louise Pratt in her 1991 essay, "Arts of the Contact Zone." She uses the term autoethnography to indicate "a text in which people undertake to describe themselves in ways that engage with representations others have made of them" (35). Her essay traces a Peruvian's seventeenth-century response to imperialist Spanish representations of native South Americans, and her goal is to posit the "contact zone" as that physical space in which people from different cultures meet and (potentially) share views that enable each to understand the other better. As educators, we need to equip our technical communication students with theoretical knowledge, rhetorical savvy, and, perhaps most important, organizational knowledge needed to demonstrate to their colleagues from other professional classes in these zones how grossly inadequate the talking handbook is. Bolstering their rhetoric should be the argument that this role is not just a personal affront; it is gross neglect of writing's potential as a learning and thinking tool—and thus a contribution to the organizational culture—as writers wield their craft.

A "Neglected Population"

The neglect of writing's potential goes hand in hand with the neglect of technical communicators in workplace settings, yet autoethnographic accounts of those settings can perhaps help lessen the neglect. At least David Hayano would have thought as much in 1979. His article "Auto-Ethnography: Paradigms, Problems, and Prospects" notes that the first appearance of the term "auto-ethnography" in 1966, in reference to Jomo Kenyatta's study of his own native Kikuyu people. At the time this study was discounted because, as a member of the culture under study, Kenyatta purportedly lacked the outsider "objectivity" necessary for collecting valid anthropological data. Hayano does not dwell on the colonialist ideology underpinning such critique but rather focuses on the prospects and possibilities of autoethnography as he sees them. One prospect that proves particularly relevant to the study of writers' subjective work identity is autoethnography's capacity to give a forum to "the voices from within—the internal political affirmation of cultural diversity and autonomy for sometimes neglected populations and peoples" (103).

His point resonates strongly with these writers, as a "neglected population" in many workplace cultures attempts to limn the terms of neglect. We often discuss these terms with respect to the epistemologies of composition as professional writers were socialized into these epistemologies as students—learning writing as the "form" to be used to embrace literature's "content," for example. We discuss these terms with respect to the positively correlating factor of gender, too. As in many other professional contexts, the work done by women is perceived as somehow lesser, is remunerated accordingly, and is institutionalized as secondary—through explicit organizational structures and practices. Lisa and Rachel focused on sexist discourse explicitly—in Lisa's case to point out ways in which a discourse of sexism subtly infiltrated even editing practices and in Rachel's case to note uneven pay for equal work.

I have used the term "discourse" in this last instance in the postmodern sense. That is, although these workplaces evidenced no overt instances of sexist statements, a sexist set of practices imbued them all the same. Gee, Hull, and Lankshear provide a helpful definition: A Discourse is composed of ways of talking, listening, reading, writing, acting, interacting, believing, valuing, and using tools and objects, in particular settings and at specific times, so as to display or to recognize a particular social identity" (10). Their text demonstrates how certain discourses do or do not mesh easily with the discourse of fast capitalism, in some cases positioning workers to accept and adhere to its tenets. They say:

> Seeing the new capitalism for what it is—namely a new Discourse in the making—allows us to juxtapose it with other, competing, overlapping, and mutually adjusting Discourses, such as a critical version of sociocultural literacy (which informed the critique in this book), the various Discourses of school reform, and a variety of other community-based and public sphere Discourses. (165)

In my analyses of autoethnographies composed in Cultures of Professional Writing, I have identified a number of discourses at work in writers' workplace cultures similar to those mentioned by Gee, Hull, and Lankshear. These discourses work subtly to position employees, and so I have identified them as "underlying discourses." The full list can be seen at http://archive.ncte.org/ccc/2/53.2/henry/D-4tbl.htm, and a few merit discussion here to illustrate the potential of autoethnographic research in workplace sites.

Both Ana and Mandy endured the double whammy of sexist and racist discourses in their cultures. In Mandy's case, she did a brilliant job of demonstrating how these discourses, by virtue of the politics of exclusion that they forwarded, resulted in a waste of time and money that undermined the organization's goals. Kolar noted how powerful political discourses on Capitol Hill positioned writers to fabricate letters from constituents. Cass, Diane, and Lois each traced specific ways in which a management discourse in their workplaces sought to manipulate. Anita, George, and Sue identified ways in which the disciplining discourse of the military influenced textual properties. In George's case, he demonstrated how "bureaucratese" resulted not from inept writers but from this disciplining discourse at work in the military's review processes, known as the "chop chain." (See also Henry and "George.")

Other effects of specific discourses on writers include frustration, low morale, and withdrawal of engagement; more general effects on the knowledge and behavior in specific workplaces include tactics deployed to work the bureaucracy and alienation of whole work units (see http://classweb.gmu.

edu/classweb/jhenry/D-5_table.htm and D-6_table.htm for the exhaustive lists). Whereas these effects might be discounted as personal and idiosyncratic, postmodern understandings prompt us to see them as social and systemic, as the result of private and public sphere discourses.

Here is where another potential of autoethnography as identified by Hayano proves apropos: "its potential advisory capabilities in programs of change or development" (103). We can map such discourses, their interrelationships, and their forms and effects in specific locales. By so doing, we construct what Jean-François Lyotard terms *petits recits*—small stories that offer ways and means of resistance to the master narratives (or discourses) of modernist ideology, and we can use the knowledge thus gleaned to change and develop our composition and technical communication practices and epistemologies.

For example, Kali's abstract for her autoethnographic report, "Sink or Swim: How Employees Stay Afloat in a Navy Contracting Company," reads as follows:

> The Navy contracting industry is volatile, unpredictable, and offers little job security or recognition to its employees, which could affect the quality of their writing. None of the participants in this study agreed absolutely on the exact correlation between lack of job security and quality of writing, but all of the employees had an internal work ethic that stimulated them to work hard for personal satisfaction. This study gives a brief explanation of the ever-shifting Navy contracting business, gives short biographies of the participants, and then discusses how the individuals perceive the effects of minimal job security on the quality of their writing.

The "lack of job security" as reflected in the play that opened this article recurs as a frequent theme across the studies and seems directly related to the Discourse of fast capitalism as Gee, Hull, and Lankshear characterize it: workers who can "learn and adapt quickly" will likely have to "adapt" to layoffs at one organization, then to seek work with another. Kali alludes to the writer psychology that such subjectivity engenders: Can one truly maintain a "quality of writing" when potential layoff is an ever-present part of one's writerly sensibilities? And what might "quality of writing" signify? By fast capitalist standards, quality writing would be writing that maximizes investors' returns on investments, and one can imagine writerly sensibility being shaped to this end, if only to maintain one's current job as long as possible. Eliminated from the equation are issues of ethics, of workers' interrelationships with colleagues, of the quality of life in the local work culture, and of the ultimate effects on other populations of the writing in which one is engaged.

Perhaps because technical communication has been perceived as that genre of composition most closely associated with potential catastrophe (witness the number of studies on the Challenger and Columbia catastrophes as at least partly rhetorical), scholars in this field have been quick to address such issues of writerly sensibility in the realm of responsibility and ethics. Thus Cynthia L. Selfe and Richard Selfe, Jr., ask us:

> To whom are technical communicators responsible? Their corporate employers? The public who uses the corporation's products? The professional societies to which they belong? Who is responsible for the information that technical communicators produce? Who controls the information technical communicators produce? Why is this so? Whose interests are being served in controlling information and in defining the various roles and responsibilities of technical communicators in the way that we now do? (330)

The status of technical communication as represented here in fact reflects the status of workplace writing more generally, and the answers to their questions, in many organizational sites, is unfortunately simple: the market, as recently reconfigured by fast capitalism.

In such a context, what lessons for theorizing and teaching technical writing from a cultural studies approach might be taken from the *petits recits* composed by professional writers such as Kali? First and foremost, it would seem, would be making writers' subjectivity as lived in organizational contexts part and parcel of any technical writing curriculum. We can offer courses such as Cultures of Professional Writing that are wholly organized around the autoethnographic project, or we can integrate short fieldwork forays into courses with other objectives. We can render service-learning courses a bit more complex by layering on autoethnographic elements that require students to analyze the organization even as they write for it, and we can frame internships similarly. At the very least, we can require our technical communication students to conduct some interviews with practicing technical communicators that broach the topics of organizational culture and writing's (and writers') places in it. If we want to hearken to the kinds of activism enjoyed by those of us who were schooled before the advent of globalization, we might also integrate some measure of "organizational reflexivity" into our syllabi, making our respective college or university structures and practices part of our material. And finally, we can also use the World Wide Web to share more *petits recits*, to the ends of imagining instruction, beliefs, and alliances that will help all writers (ourselves included) better grasp writing's representations and uses in workplace cultures.

In her "Introduction" to *Autoethnography: Rewriting the Self and the Social*, Deborah E. Reed-Danahay says that "[t]he ability to transcend everyday conceptions of selfhood and social life is related to the ability to write or do autoethnography. This is a postmodern condition. It involves a rewriting of the self and the social" (4). In light of the Discourses shaping the new work order and writers' lives in workplace cultures as reflected in the *petits recits* of the Cultures of Professional Writing research project, it seems vital that we devise ways to enable students to transcend everyday conceptions of selfhood and work life, for these conceptions ultimately disempower them, in the process masking the huge powers of discourse in workplace cultures. The concluding section sketches some paths.

Rewriting Workplace Cultures

Carl Herndl's point that ethnographic research on workplace cultures should remain vigilant of dominant discourses and seek to revise them has needed no reiteration as researchers each spring take to investigating their workplace cultures. Enduring above all a discourse that positions their expertise as marginal in the organization's life, writers yearn for the means to find other organizations where such is not the case or to somehow change local views that see writing as a simple act of communicative packaging. As lone "agents" in a local culture, they stand little chance of doing so, particularly given their low organizational status. This predicament underscores the need for alliances between scholars and writers in the workplace, the better to build other archives of organizational structures and practices and the better to foster border crossings that will infuse productive revision of discourses in each.

One discourse in need of revision in the academy is that of writing administration: when Sharon Crowley identifies as problematic to composition the institutional structures through which it is administered, I think of a discourse of scientific management that calculates university revenues partly on the basis of large numbers of students who can be taught in composition and technical communication courses with modest expenditures. The result are courses taught mostly by part-time faculty who must hustle to make ends meet and who simply cannot have the time necessary to join their students in meaningful research projects that inquire into the workings of discourse in organizational settings—those settings where most students will exercise most of their discursive talents for the rest of their lives. This argument is familiar to compositionists, I know, but in light of the above discussion, it bears repeating. As for technical communication, its status varies dramatically according to local structures, no doubt, but at most colleges and universities, it remains a service course. In my former English department, for example, the course was often assigned to adjunct instructors who do no research in technical communication and who have virtually no practical experience as technical communicators! To revise workplace cultures, we must revise the administrative structures that sap writing of its potential for inquiry into the workings of discourse in all kinds of sites.

Another discourse in need of revision is that of technical communication's own epistemologies, its own theories of the nature of knowledge and more explicitly of the nature of technical communication's content. When queried as to that "content," I think most instructors would harken to issues of craft—rhetorical strategies, organizational principles, "style," grammatical soundness, editorial rigor, and the like. When I am in a modernist mode of thinking, I certainly harken to such notions, and I do so with conviction as to their value in aiding writers to augment their writerly sensibilities. But when I supplement this view with one informed by postmodern inquiry, particularly as informed by empirical study of the workings of discourses in organizational cultures, I see another content for technical communication: the construction and mediation of realities through language. To reiterate a premise stated earlier: language is the very material from which realities are socially constructed. We are born into language and we can know realities only through the categories and relationships that language has provided. Through language practices, discourses create subject positions that exert enormous influence over how we perceive our lives and how we live them. When we write, we engage with these discourses, strongly subjugated by certain among them, more liberally positioned by others, to articulate positions that in their turn contribute to discursive scenes which may or may not bear on others. Surely we must teach this content of technical communication, too.

These two discourses in need of revision go hand in hand: To embrace an epistemology that acknowledges communication's content as the construction and mediation of realities through language, we will need whole programs that focus on writing and communication, rather than departments that offer composition or technical communication under the paradigm of "service." Sharon Crowley has proposed such a curriculum, a curriculum

> that examines composing both in general and as it takes place in specific rhetorical situations such as workplaces and community decision making. While I can envision challenging courses in invention or style or argumentation being offered in such a curriculum, I would hope that such a course of study would not confine students to practice in composing. Rather, it would help them to understand what composing is and to articulate the role it plays in shaping their intellectual lives. The topmost reaches of an under-

> graduate curriculum in composing would study histories of writing, debate the politics of literacy, and investigate the specialized composing tactics and rhetorics that have evolved in disciplines, professions, civic groups, women's organizations, social movements, and political parties—to name only a few sites where such investigations could fruitfully take place. (262)

Crowley's concerns are specific to composition, but one can easily imagine the significant role that technical communication would take in such a curriculum. Enacting this curriculum would in fact represent something *akin* to the movement in English departments in the last decade toward cultural studies. While this movement has been motivated by literature scholars' desire to address their central problematic more fully, it also represents a will to power grounded in a conviction that to interpret texts adequately in the twenty-first century, students must learn to interpret in specific cultural contexts. And because the definition of "texts" now includes just about any discursive representation, the move toward cultural studies represents a will for a master discipline.

This will makes perfect sense, given the demise of the humanities in the waning decades of the last century. Particularly in an age of technosupremacy and its alliance with many of the discourses of fast capitalism—discourses that would position new graduates as "information workers"—a strong resistance within the academy is imperative. But the failing of cultural studies, I believe, has been its relentless insistence on forming students as critical discursive consumers all the while wholly ignoring their formation as critical discursive producers in any genre other than the academic essay. We need to position students as something such as "discourse workers," with all the attendant implications for curricula sketched above. We need to foster in our students an aptitude for cultural criticism, grounded in skills at cultural analysis in real settings. With this kind of knowledge and skill, students can perhaps begin to see the cultural analysis that they conduct in other courses as dovetailing directly with their roles in cultural production and reproduction in workplace settings and to seek with us a means to influence the policies of those settings.

Such an approach to technical communication is not just epistemologically sound; in the era of globalization it is ethically imperative—with respect to our students and with respect to our discipline. The Bureau of Labor statistics cited earlier paint a rosy picture for technical communicators—28,000 new job openings due to growth and net replacements by 2012 (emptab 21). But these projections, notes Katherine Boo, "rely on corporations to link their domestic downsizing to work they now send abroad—a connection that some corporate leaders are loath to make" (59). As "outsourcing" grows as a strategy used by management to maximize profits, domestic jobs such as technical communication could well plummet in the years to come. Boo states: "Other analyses suggest that the number of American jobs lost to this phenomenon will soon reach a million, as the Indian and Chinese back-office sectors expand by thirty percent a year" (59). One such analysis is that of Jaffee, Bardhan, and Kroll, who in *Globalization and a High-Tech Economy: California, the U.S. and Beyond*, confirm the growing outsourcing trend particularly in the high-tech sector. Within that sector, as long as technical communication is perceived as mere transcription of a kind and as long as our technical communicators are kept in the bottom boxes of organizational charts, the outsourcing of this writing looms inevitable. If we don't alert our students to this trend, we are nothing short of disingenuous.

As for our discipline, compare it with those of other professionals in the workplace—management, for example. In my former Cultures of Professional Writing course, I continued to use as a reading John Van Maanen's "The Smile Factory: Work at Disneyland," even though the article dates from authoethnographic fieldwork during the 1960s. Students enjoyed the "retro" view on much of Disney mantra that now circulates quite freely in mass culture depictions, along with Van Maanen's ability to cast everyday observations in sociological terms. I also included Edgar Schein's "Organizational Culture," even though this article dates from 1990, because I wanted students to see how an organizational psychologist approaches workplace analysis and I wanted them to know that organizational psychology has always aligned itself, as a field, with management studies. Both Van Maanen and Schein teach at MIT's Sloan School of Management, regularly instructing their charges in organizational aptitude that they will wield when they move into the boxes at the tops of organizational charts—this, while our technical communication students will be filling those boxes at the bottom. As professors of technical communication, aren't we ethically obliged to offer our students some similar footing in organizational culture analysis?

WORKS CITED

Bauman, Zygmunt. *Globalization: The Human Consequences*. New York: Columbia UP, 1998.

Berlin, James. *Rhetorics, Poetics, and Cultures: Refiguring College English Studies*. Urbana: NCTE, 1996.

Boo, Katherine. "The Best Job in Town: The Americanization of Chennai." *The New Yorker* 5 July 2004. 54–69.

Bureau of Labor Statistics. Table 2: Employment by occupation, 1996 and projected 2006. 4 July 1999. http://www.fedstats.gov/index20.html.

———. Table 11: Employed persons by detailed occupation, sex, race, and Hispanic origin. 4 July 1999. ftp://ftp.bls.gov/pub/special.requests/lf/aat11.txt.

———. Employment by occupation, 2002 and projected 2012. 10 July 2004. http://www.bls.gov/emp/emptab21.htm.

Crowley, Sharon. *Composition in the University: Historical and Polemical Essays.* Pittsburgh: U of Pittsburgh P, 1998.

Debs, Mary Beth. "Corporate Authority: Sponsoring Rhetorical Practice." *Writing in the Workplace: New Research Perspectives.* Ed. Rachel Spilka. Carbondale: Southern Illinois UP, 1993. 158–170.

Dias, Patrick, Aviva Freedman, Peter Medway, and Anthony Paré. *Worlds Apart: Acting and Writing in Academic and Workplace Contexts.* Mahwah, NJ: Lawrence Erlbaum, 1999.

Duin, Ann Hill, and Craig J. Hansen, eds. *Nonacademic Writing: Social Theory and Technology.* Mahwah, NJ: Lawrence Erlbaum, 1996.

Faigley, Lester. *Fragments of Rationality: Postmodernity and the Subject of Composition.* Pittsburgh: U of Pittsburgh P, 1992.

Garay, Mary Sue, and Stephen A. Bernhardt, eds. *Expanding Literacies: English Teaching and the New Workplace.* Albany: SUNY P, 1998.

Gee, James Paul, Glynda Hull, and Colin Lankshear. *The New Work Order: Behind the Language of the New Capitalism.* Boulder, CO: Westview Press, 1996.

Harvey, David. *The Condition of Postmodernity.* London: Blackwell, 1989.

Hayano, David M. "Auto-Ethnography: Paradigms, Problems, and Prospects." *Human Organization* 38 (1979): 99–104.

Henry, Jim. *Writing Workplace Cultures: An Archaeology of Professional Writing.* Carbondale and Edwardsville: Southern Illinois UP, 2000.

Henry, Jim, and "George." Workplace Ghostwriting. *Journal of Business and Technical Communication* 9 (1995): 424–45.

Herndl, Carl G. "Teaching Discourse and Reproducing Culture: A Critique of Research and Pedagogy in Professional and Non-Academic Writing." *College Composition and Communication* 44 (1993): 349–363.

Hull, Glynda, ed. *Changing Work, Changing Workers: Critical Perspectives on Language, Literacy, and Skills.* Albany, NY: SUNY P, 1997.

Jaffee, Dwight, Ashok Deo Bardhan, and Cynthia Kroll. *Globalization and a High-Tech Economy: California, the US and Beyond.* New York: Kluwer Academic Press, 2003.

Jameson, Fredric. *Postmodernism, or, The Cultural Logic of Late Capitalism.* Durham: Duke UP, 1991.

Johnson-Eilola, Johndan. *Nostalgic Angels: Rearticulating Hypertext Writing.* Norwood, NJ: Ablex, 1997.

Lay, Mary M. and William M. Karis, eds. *Collaborative Writing in Industry: Investigations in Theory and Practice.* Amityville, NY: Baywood, 1991.

Lionnet, Françoise. "Autoethnography: The An-Archic Style of Dust Tracks on a Road." *Reading Black, Reading Feminist.* Ed. Henry Louis Gates. New York: Meridian Press, 1990. 382–414.

Lyotard, Jean-François. *The Postmodern Condition: A Report on Knowledge.* Trans. Geoff Bennington and Brian Massumi. Minneapolis: U of Minnesota P, 1984.

McBeth, Sally. "Myths of Objectivity and the Collaborative Process in Life History Research." *When They Read What We Write: The Politics of Ethnography.* Ed. Caroline B. Brettell. Westport, CN: Bergin & Garvey, 1993. 145–162.

Miller, Susan. *Textual Carnivals: The Politics of Composition.* Carbondale, IL: Southern Illinois UP, 1991.

Odell, Lee, and Dixie Goswami, ed. *Writing in Nonacademic Settings.* New York: Guilford, 1985.

Pratt, Mary Louise. "Arts of the Contact Zone." *Profession 91.* New York: MLA, 1991. 33–40.

Reed-Danahy, Deborah E. "Introduction." *Auto/Ethnography: Rewriting the Self and the Social.* Ed. Deborah E. Reed-Danahy. New York: Oxford UP, 1997. 1–19.

Reich, Robert. *The Work of Nations.* New York: Vintage, 1992.

Russell, David R. *Writing in the Academic Disciplines, 1870-1990: A Curricular History.* Carbondale: Southern Illinois UP, 1991.

Schein, Edgar H. "Organizational Culture." *American Psychologist* 45.2 (February 1990). 109–119.

Selfe, Cynthia L., and Richard L. Selfe, Jr. "Writing as Democratic Social Action in a Technological World: Politicizing and Inhabiting Virtual Landscapes." *Nonacademic Writing: Social Theory and Technology.* Eds. Ann Hill Duin and Craig J. Hansen. Mahwah, NJ: Lawrence Erlbaum. 325–358

Slack, Jennifer Darryl, David James Miller, and Jeffrey Doak. "The Technical Communicator as Author: Meaning, Power, Authority." *Journal of Business and Technical Communication* 7 (1993): 12–36.

Spilka, Rachel, ed. *Writing in the Workplace: New Research Perspectives.* Carbondale: Southern Illinois UP, 1993.

Taylor, Frederick W. *The Principles of Scientific Management.* New York: Harper & Brothers, 1911.

Van Maanen, John. "The Smile Factory: Work at Disneyland." *Reframing Organizational Culture.* Ed. Peter J. Frost, Larry F. Moore, Meryl Reis Louis, Craig C. Lundberg, and Joanne Martin. Newbury Park, CA: SAGE, 1991. 58–76.

Virilio, Paul. *Speed and Politics: An Essay on Domology.* Trans. M. Polizzotti. New York: Semiotext(e), 1986.

Yates, JoAnn. *Control through Communication: The Rise of System in American Management.* Baltimore: Johns Hopkins UP, 1989.

Chapter Nine

Rhetoric as Productive Technology

Cultural Studies in/as Technical Communication Methodology

Michael J. Salvo

> Power in a democracy must remain in the hands of common citizens, and citizens must finally decide on the courses that government and business are to take. While the voice of experts must be heard, the people must choose the heading the community should follow.
>
> —James Berlin, *Rhetorics, Poetics and Cultures*

James Berlin's argument in *Rhetorics, Poetics and Cultures* names race, gender, and class as the three dominant matrixes that will shape cultural studies inquiry, particularly in literary studies (33). It is telling that technology remains unmentioned in Berlin's mapping of cultural studies, for even though *Rhetorics, Poetics and Cultures* carries a publication date of 1996, the book was largely complete before Berlin's untimely death in 1994. At that time, the Web as we know it had not yet become the ubiquitous communications environment it has become, and the economic boom of the 1990s was just beginning. Mosaic had not yet become Netscape, and Microsoft had not yet recognized the threat the Web posed to its business model. MP3 files were an as-yet unknown cultural phenomenon, limited to their original role as a highly compressed file format for sharing video and sound files. Computers and writing studies were making a transition from studying stand-alone computers to seriously considering the impact of local area networking to pedagogical practice. Digitally based desktop publishing was challenging traditional labor-intensive modes of publishing, challenging assumptions about the roles of technical writers. Boundaries between technology and other cultural elements were blurring but were not yet recognized as having morphed into technoculture. And so it is that this chapter imagines what Berlin may have said about the World Wide Web.

Although Berlin's definition of postmodern critical literacy omits technology, meaningful action is not possible without accounting for technology: "In learning to gain at least some control over [textual forms], students become active agents of social and political change, learning that the world has been made and can thus be remade to serve more justly the interests of democratic society" (112).

Indeed, the same argument regarding action and agency persists in the literary/rhetorical divide that existed before the technological turn: just as literary studies aestheticizes texts it did not produce, technocultural studies (for the most part) comments on, criticizes, and aestheticizes a technoculture that it remains apart from.[1]

Rhetoric and its teachers have sought to enable rhetorical agents to participate both in the texts that accompany technocultural production as well as the production of the artifacts of technoculture. Therefore, this essay seeks to revise Berlin's definition of postmodern critical literacy to include technological development, the artifacts it produces, and the culture those artifacts are deployed within.

> Learning to gain some control over communication forms and the technologies that enable them, students become active agents of social, political and technical change, learning that social and technological worlds have been made and can thus be remade to serve the interests of democratic society.

Interestingly, critical technical communication studies have already begun describing discourses that reveal active participation by communicators in the design and deployment of technology (Johnson; Salvo), and some in our field have begun using cultural studies methods to develop new approaches to research and curricula (Longo, 1998, 2000; Herndl, 2004). Beyond the field of professional and technical writing, similar engagements at the crossroads of technology and culture appear variously labeled *A Critical Theory of Technology* (Feenberg), *Visual Intelligence* (Barry), and "electracy" (Ulmer), while emerging studies of information management, such as knowledge management (Wick), information architecture (Wurman), and information design (Jacobson) are working toward functional and practical means of enabling human organization and mapping of the swelling tides of data and information produced with digital technology. The challenge remains for allowing cross-fertilization, toward enabling hybridity, across the barrier between analysis and engagement.

Each discourse has a specific history and population of practitioners, but bringing these various titles and names together reveals shared discursive, ideological, and cultural goals. My list is meant to be representative rather than exhaustive. The aim is to highlight the active, cultural engagement they each foreground and which contrasts these discourses from passive aestheticization, reflecting a long-running discussion in English studies that Miller takes up in her posthumous review of *Rhetorics, Poetics and Cultures.* I am not claiming here that cultural studies methodologies necessarily seek passivity or do not value action. Rather, I assert that cultural studies does its most insightful work in the analytic phase by mapping discourses, institutions, and flows of power on a virtual map of culture. This productive analytic thrust of cultural studies can effectively inform action through critically examining design, mapping the discourses that inform design, as well as revealing the complex networks of power and the interests that are served and subsumed in different designs.

Late in Grossberg, Nelson and Treichler's collection *Cultural Studies,* Angela McRobbie wonders about the goal of cultural studies, asking simply "Why do it?" Concerned that academic cultural studies stops with analysis rather than enables action, she questions the purpose and commitment of cultural studies. McRobbie continues: "There is relatively little direct engagement with the role of cultural intellectuals . . . in any of the emergent global socio-political formations of the 1990s" (721). For all of the attempts to create a cultural studies praxis, McRobbie's observation that there is little "direct engagement" by intellectuals remains striking. There is much discussion of what is happening and what others are doing in and with technology, but no models for engaging technologies and the discourses surrounding their design, development, and deployment in culture (see Feenberg for a lonely exception). Bringing the powerful analytic and descriptive methodologies from cultural studies into technical communication and using them to inform the active, engaged, and productive elements of technological invention and design most specifically offers technical communicators an effective means for engaging political, social, and discursive implications of technoculture. And, perhaps more directly relevant to this chapter and collection, wedding cultural studies analysis to design offers rhetoricians an opportunity to demonstrate language in action and showcase rhetorical agency, which clearly serves the interests of rhetoricians. But providing this example of cultural studies as the analytic engine of technocultural inquiry also offers one example in response to McRobbie's question of "Why do it?"

Put another way, cultural studies is most itself when heretical, and therefore this argument is part of the antitradition of the discourse(s) of cultural studies which, following McRobbie, questions the field's own basic assumptions and values regarding not only what cultural studies is, but also what it might be and what it might become: "Cultural studies these days embraces an astonishing number of people and positions (no one has managed to register the trademark). . . . Whichever wing of cultural studies we examine, it must be remembered that they represent loose, shifting and occasional coalitions of scholars" (Carey 2). Although cultural studies may be comprised of "loose, shifting and occasional coalitions of scholars," these scholars similarly study artifacts of culture by placing these artifacts into the context of their production in order to account for (in Althusser's construction) the "reproduction of the conditions for production" or (in Hall's "Encoding, Decoding") the hegemony in meaning making that is "naturalized."

Many strands of cultural studies are also action oriented. Tony Bennett's "Putting Policy into Cultural Studies," which appears in Grossberg, Nelson, and Treichler's collection along with McRobbie, makes one of the most powerful arguments for moving to action and engaging with culture. McQuail similarly calls for cultural studies to:

> broaden its scope to include the larger framework in which cultural issues are problemetized and in which varieties of culture are defined and ordered in a social process of production and distribution (as well as reception). The analysis of culture is not an end in itself, but it is a necessary component in formulating policy and it is a task in which policy-makers are not especially qualified to undertake. (52)

Cultural critics and theorists can participate in the construction of culture, McQuail asserts, and can reform cultural interrelations according to the values held by cultural studies, even if these values are only "loose, shifting and occasional." It remains the challenge of cultural studies to develop a place from which to suggest formulations, a ground on which to proclaim a preference for one design over another, to propose "policy" as called for in McQuail's essay.

One problem with McQuail's call for policy participation, however, is a misunderstanding or misrepresentation of the very power relations that define communication between producers and consumers in broadcast culture. Policy is created by production insiders to enhance the efficiency and profitability of communication (Johnson-Eilola), and cultural studies generally comments on rather than participates in production of this communication. Cultural studies has certainly expressed desire to participate in the production and content of broadcast images and sound, but has been unable to effectively participate in the making of broadcast media. In this context, then, cultural studies discourse has concentrated on consumption and leisure rather than participation and sites of work (Garnham 60).

Here is an opportunity not only for technical communication to understand itself as both part of humanities curricula and a site capable of informing the larger project of cultural studies. Focused as it is on the workplace, technical and professional communication is uniquely able to provide "policy" insights not just for the better use of leisure time but to inform postindustrial labor policy as it understands the discourses of workplace production, of interrelations of individuals, values, ideologies, and the discourses they produce. Cultural studies methodology not only brings insight and value to the cultural discourses it engages, but also works toward greater engagement with dominant culture, not in the form of selling out or of giving up, but in making cultural formations more responsive to a wider range of concerns and in blurring the boundaries between producers and consumers.

The producer/consumer binary is something I have addressed elsewhere (Salvo, "Ethics") but which bears mention here insofar as culture is moving from communication models based on few-to-many in broadcast versus many-to-many as well as the production of microaudiences and niche markets that can best be described as few-to-few and many-to-few. Peer-to-peer software here is a clear example of the rise of many-to-many broadcasting: anyone with the appropriate software can make files available to anyone else with appropriate software. The media of distribution is open to all with digital communication equipment and an Internet service provider, much to the chagrin of the Recording Industry Association of America (RIAA), which has recently resorted to increasingly invasive methods of detecting breaches of intellectual property law. Indeed, recent actions of the RIAA and others have been attempts to criminalize not just violations of copyright and intellectual property, but of peer-to-peer file sharing more broadly. In effect, the broadcasters are claiming that only their transmissions are to be afforded cultural capital (Schwartz).

However, this chapter is not about recent developments in intellectual property or in peer-to-peer networking but about what technical and professional communication can make of cultural studies research as well as what the field has to offer the discourse of cultural studies. It is about advancing technical communication research to both participate in cultural studies discourse and to move that discourse forward, to add technical communication's engagement of workplace discourse and power dynamics to cultural studies while also enriching technical communication.

This chapter defines technology and its deployment in culture as technoculture, and technical communication as the communication that accompanies the invention, design, arrangement, and deployment of technology in culture. Employed in technical communication pedagogy, cultural studies methods can inform efforts to create not only critical but also active rhetorical agents—those communications experts whom Robert Johnson calls technical rhetoricians in his book *User-Centered Technology*. Educating such technocultural agents is not limited to analysis and critique, but moves toward rhetorical action. In this definition of technical communication, cultural studies is not an inert critical positioning of the technorhetorical gaze, but a mode of informing and sanctioning critical action.

Ultimately, the role of the expert is shifting in culture at the same time that the role of the academic is shifting. Experts are being asked by citizens to consult with communities to create meaningful, usable, intelligent solutions that are built with attention to local practices, constraints and economic boundaries, from building *for* users to building *with* users. This shift parallels the shift Berlin wrote about throughout his career in which he encouraged his colleagues in English and in the humanities generally to engage the culture they commented on, to enact cultural design as the outcome of cultural study, to help students become active rhetorical agents as they performed critical rhetorical analysis. In other words, rhetoric is an active production of meaning rather than passive interpretation of existing texts. In her critique of Berlin's posthumously published *Rhetorics, Poetics and Cultures*, Susan Miller even more forcefully argues for an action-oriented rhetorical pedagogy:

> for enabling students to act through language, first by placing its differential modes of making in the center of our teaching, as Jim [Berlin] suggested. By teaching texts rather than their making, by teaching awareness rather than rhetoric, and by teaching the power of meaning rather than the making of statements, we inadvertently reproduce a politics that is aware but passive. Rhetoric is not, that is, semiotics. And while it often suits us to equate the two . . . writing is not reading. (209)

Following Miller's distinctions, my formulation of cultural studies calls for a *productive* rhetoric attuned to the demands of technoculture, where technology and communication are brought together in unique and demanding ways. The production of effective arguments needs to be the basis of rhetorical education. With the ubiquity of the World Wide Web, the means to engage in ongoing discourses is open to both cultural critics and to technocultural workers, as well as to teachers preparing students to both work in the age of information and participate as democratic citizens.

The next section of this chapter examines five technical communication texts and the developing field of user-centered design to illustrate the cultural development of a new kind of postmodern expert. This expert is asked to apply, to act on, the knowledge learned through cultural studies methods to influence the development of culture: to become the "policy experts" described by McQuail. Experts in technical communication would not only describe the cultures they study but would actively engage these formations and design alternative models of interrelations among humans, technologies, and the institutions that support them, not with the aim of increasing efficiency but of increasing opportunities for human agency and engagement—of increasing democratic engagement. Democracy is defined here as a place where dialogue has an opportunity to impact the design and deployment of technologies in culture and of the construction and regulation of relationships among competing groups within culture. The solution is not structural—that is, there is no privileged place from which the expert can make suggestions that guarantees their successful adaptation and application. Rather, the solution is discursive, and the postmodern expert has the added responsibility of helping educate and prepare those interested and invested in the solution to be able to effectively engage dominant exercises of power. Perhaps most valuable is the ability to recognize when discourse has the potential to change outcomes.

These five recently published texts in technical and professional communication gesture towards reformulating the professional rhetorical agent as I have been describing: *Community Action and Organizational Change: Image, Narrative, Identity* by Brenton Faber; *Writing Workplace Cultures* by Jim Henry; *Power and Authority in Technical Communication*, volume one of a two-volume collection edited by Teresa Kynell-Hunt and Gerald Savage; *The Rhetoric of Risk* by Beverly Sauer; and *Rhetoric as an Art of Design* by David Kaufer and Brian Butler. In the following sections, I'll use these texts as points of departure for and, in Sauer's case, as an example of my larger argument.

CHANGE AS CONSTANT: LEARNING TO LEARN

> "... academics have resisted the role of an expert and instead have found productive ways to work with groups and organizations in their communities." (Faber 180–181)

Faber's *Community Action and Organizational Change* offers one professional communication consultant's experience at a variety of workplaces and articulates a new role for the expert in postindustrial culture. The new professional, according to Faber, is a freelance agent asked to play the role of consultant who suggests alternative structures that the local community must buy into and enact. This is a very different role to play than the modernist professional, a professional who, set apart from local communities, dreamed up decontextualized images of utopian relationships that were written over existing discourse relationships. Rather than dismissing or erasing the local, the postmodern professional acts in concert with, and becomes enmeshed in, local culture.

I suggest that, rather than "give up" the role of expert, we (academics and consultants) are instead redefining the meaning of "expert." An expert, in the narratives that create Faber's book, is not someone who knows better than the local culture. The paternalism Faber locates in many experts' tone and attitude is absent from this new definition of expert. Indeed, in Faber's reconstruction, the expert becomes someone who works with and among local populations, and often uses the word *partner* to describe the new role. Rather than engaging these communities from above, as experts telling local populations how to better address their own needs, Faber's reconfigured expert is a partner whose first responsibility is to learn the local language, customs, and stories. As such, an expert should become one of the tribe and should make contextualized suggestions that are respectful of and comprehendible within the established working relationships of the community, a role parallel to that proposed in contemporary rearticulations of ethnographic methods in cultural anthropology.

Being expert in communication, in technical writing, in usability, does not exclude one from playing the role of expert in an accompanying technical endeavor. However, it is important to recognize both the opportunities for intervention as well as the limits of that expertise. The expert's role then becomes a self-conscious analysis and comparison of the local conditions with the previous experience and knowledge of the expert. And it is also where invention (both in the rhetorical and the technological sense) becomes important: given this context, these narratives, and a unique population, what characteristics will a solution have to have? Additionally, the expert cannot deploy an existing solution created in another context. Rather, the analysis becomes one of critical comparison: which parts (or modules) of previous solutions can be redeployed here, and which elements need to be fashioned anew? And in many cases, it is the interface between different systems that needs to be most closely redesigned and redeployed.

I am not suggesting here that technical and professional communicators develop new expertise for which we are neither prepared nor well-suited. However, technical and professional writers' existing expertise in effective communication, coupled with the role of user advocate, informed by cultural studies analysis, can and should allow practitioners and academics to contribute to the invention of new technologies. Technical rhetoricians have the unique capacity to contextualize combinations of technologies and social relations so that local communities and stakeholders can transform them. And this process is not just about taking back control from technical "experts," but about returning the sense of everyday, ethical engagement with the very people and communities with whom we work. We are not displacing our partners from their work, but recontextualizing their understanding of the work they do, helping them gain a sense of local control and engagement and demonstrating how to do better the things they already do well. Opportunities for local intervention have been weakened by the processes and *narrative* of globalization, but it is only when the expert listens to local narratives and understands how the community represents itself in relation to larger cultural narratives that interfaces respecting local cultures can be constructed.

Interestingly, this definition of expert is aligned with democratic cultural practices. In democracy, every citizen is the local unit of political and social power and is responsible for decisions that affect his or her worklife. At the same time, each member has a distinct expertise, and these fields articulated together (both in terms of connection and narrative) create community. Locally, I rely on my dry cleaner's expertise to remove the spots I inadvertently get on my clothes. But I would not have the same respect for advice he would offer on my plans for a backyard deck. The local carpenter would have my full attention in that matter, particularly if I am looking at mass-market plans and she is offering advice on local conditions: "Winter is rough so I'd add a foot to the pilings and use 2-by-6 instead of 2-by-4." But then that begs the question: What is the technical communicator's expertise?

Articulating Expertise: Archeology of Work

Jim Henry's *Writing Workplace Cultures* offers a starting place for professional writers to begin articulating the complex webs of discursive networks that make up any workplace. The book collects narratives written from 1993 to 1996 at a variety of educational institutions and worksites, from corporations and professional associations to government agencies and nonprofits. Perhaps the most striking observation of Henry's work has been made in Dale Cyphert's review, which appeared in the *Review of Communication* in 2002:

> Then we realize we are engaged in a dialogue with the archeologist. Henry's book is not "about" workplace writing. The dig is instead an exploration of the relationships between document creation and organizational culture, between literacy and orality, between community and critique, between individual and collective identity, between rhetorical performance and rhetorical text. The real treasure unearthed by Henry's archeology is material evidence of conflicts and controversies at the heart of contemporary rhetorical theory. (26)

Cyphert's observation that *Writing Workplace Cultures* reveals much about contemporary rhetorical theory is certainly accurate, yet Cyphert and the students to whom I have assigned the book have missed an important facet of the text. It is not a book that rewards cover-to-cover reading. Rather, it is best read as a reference text, skimmed and consulted in order to understand the peculiar goings-on—the contexts, conflicts, and discourses—that support and sustain professional writers' roles at a particular worksite. It is a new kind of reference text for a new kind of professional.

The charts and diagrams which attest to the sheer variety of the workplaces Henry's students have excavated are invaluable resources for writers faced with the questions of "What do I do?" and "Where do I start?" Faced with such a daunting challenge, Henry's text encourages students to analyze their institution and their position(s) within it, and then begin to unpack its discursive workings. Encouraged by their teachers to be active rhetorical agents, goaded by counterhegemonic discourses in their coursework, and taught postmodern mapping strategies as part of their rhetorical training, many young writers express dismay at being thrust into workplaces where their voice is subsumed into the cacophony of discourse. Understanding the scope of the change they are capable of enacting is perhaps one of the most important lessons new workers will need to learn. First, new workers have to become part of the team they are joining, to understand the rules by which it operates, to know its goals, practices and procedures. Henry's archeological structure and lexicon is helpful here, casting young writers as students of the new workplace cultures of fast capitalism, so different and yet so dependent on the same elements that drove the cultures from which they have arrived. Like Faber's reconfigured expert, newly credentialed writers are only trusted at their new workplaces after they have become effective team members, after they have effectively learned local customs, become fluent in local discourses, and shared in workplace rituals.

Henry's *Writing Workplace Cultures* is therefore an effective supporting text that helps practicing professional writers articulate and critique the existing institutional cultures and discourses that surround them. Henry's text offers a starting point, particularly for writers new to their responsibilities and institutions, but does not clearly suggest particular directions for action. But before they act, technical rhetors must know the history both of their positions and of their institutions. Kynell-Hunt and Savage offer a complementary historical view of the authority invested (or, more accurately, *not* invested) in writing professionals.

BE CAREFUL WHAT YOU WISH FOR: THE POSTMODERN PROFESSIONAL

Kynell-Hunt and Savage's *Power and Authority in Technical Communication* is a collection of essays that delves into issues of the technical communicator's expertise. The collection is best understood as a history of the status and standing of writing professionals, of technical rhetors, in the workplace. The volume concludes with Faber and Johnson-Eilola's study of the changing nature of professionalism in the postmodern world. Yet the balance of the book is spent describing existing relationships as revealed through historical studies, and can best be used in comparison both with Henry's recent archeological excavations of writers in the workplace and the narratives that Henry's text encourages readers to create.

By centering the collection on the authorial voice of the technical communicator—and representing Slack, Miller and Doak's "The Technical Communicator as Author: Meaning, Power, Authority" on its tenth anniversary of publication—the editors are surveying the discourse of authorship that sheds light on issues of authority. The collection reconsiders the discourse of technorhetorical authority and recontextualizes Slack, Miller, and Doak with other voices that build on their argument. Each essay attempts a different rhetorical approach to authorizing technical discourse, yet each recognizes the limits of rhetoric to make the world in its own image. Slack, Miller, and Doak articulate this phenomenon of rhetoric: "The success of our attempts at rearticulating identities, whether purposeful or not, depends upon the tenacity of the various articulations that constitute it at any particular conjuncture" (183). The problem is articulated by Berlin as the place of *transactional* rhetoric—words that shape the conditions of interpreting both what is real and what is possible.

Understanding power and authority and the discourses that narrate them are not the same as controlling these language games. Indeed, often rhetors have to learn to recognize when situations are not rhetorical at all, but are immune to the effects or interventions of rhetoric, of language games. Instead, rhetors must learn when, where, and how they can engage with and intervene within discursive exchanges to make a difference.

Such is the limit of rhetoric—that there are indeed rhetorical situations and then there are extrarhetorical or nonrhetorical moments, times when language and symbol manipulation alone will not yield to change or alteration in the space of the community, when one discourse or technology or practice will not change no matter how articulate and well wrought the language brought to bear. The future of expertise within culture has both rhetorical and nonrhetorical aspects, and as Kynell-Hunt and Savage's collection presents the issues, we can indeed damage our rhetorical case by not recognizing the latter: by insisting on premature answers, we can choose wrong paths, and by insisting on solely rhetorical solutions to extrarhetorical situations, we can talk ourselves out of relevance. Again, Slack, Miller, and Doak offer a striking example:

> We heard recently of an industry recruiter who—venting some frustration over graduates knowing more theory that was good for

> them on the job—said, "We want robots!" This frustration has, we submit, several sources. First, and most obvious, we take this to be a plea for technical communicators to do their transmission function well. (187)

Kairos, the ability to select the right time and measure of language, is a valuable rhetorical skill that will help practitioners recognize when they have an opportunity to engage in professionalization and status-seeking, and when deadline pressures require attention to the task. Slack, Miller, and Doak argue that technical communicators have to attend to both the functional and philosophical aspects of language, or in the author's terms, for transmission and translation as well as articulation. Expertise is not limited to critical analysis but also requires action, both critical status-raising rhetorical action and more mundane functional language work. Knowing when to engage in, or recognizing opportunities for, strategic action (see de Certeau) is a rhetorical skill requiring action based on hard-won knowledge of the institution, the intensification and slacking of workflow, and the potential for language to enable change.

Although Henry and Kynell-Hunt and Savage offer technical and professional communication students and communicators useful starting points for critiquing their roles in workplace cultures, they stop short of suggesting specific strategies for enacting the role of the postmodern expert. Sauer and Kaufer and Butler do offer such strategies that can be supported by the information collected through Henry's archeological methods and understood in light of Kynell-Hunt and Savage's historical perspective.

REGULATING THE UNKNOWN: CHALLENGES TO REGULATORY WRITING

Sauer's *Rhetoric of Risk* offers reconfigurations of expertise in the context of high-risk communications and regulatory writing. The sites for her study are mines on three continents: in the United States, U.K., and South Africa. In actively engaging miners as both a source and beneficiary of technical knowledge, Sauer's work troubles the boundary between producers and consumers of technical information. The book rearticulates expertise as sets of responsibilities and obligations, and asserts that managers have an obligation to strive toward making dangerous workplaces safer for miners and to do everything possible to avoid both loss of property and loss of life. Sauer teaches us that workers have much to provide in the process of identifying and forewarning peers and power-holders of potential dangers. Technical and professional writers are uniquely positioned to bridge these populations, providing much-needed information to decision makers while valuing the insights of labor's unique perspective of the worksite.

Sauer's study brings danger to the forefront and challenges technical rhetoricians to rethink their placement as detached observers. In such situations, the boundary between "mere" words and "real" danger blurs, and direct outcomes from poorly written or mistaken documentation have real consequences in the lives and bodies of mine workers. The effects of rhetoric on high-technology and similar postindustrial commercial processes are equally real if less clearly connected to the processes they describe. Often, the effects are deferred into the future—nothing so dramatic as a cave-in results from poorly constructed regulatory standards—yet there are real consequences: promising firms that fold because their poorly documented products could not compete in the marketplace, or avoidable wastes of time and money as users are unable to complete simple tasks because documentation fails. Worse still: high-wage jobs are outsourced or "offshored" in the misperception that professional writing is mere translation of information and delivery of facts from those who know to those who lack understanding.

Sauer presents compelling descriptions of various new sources of information in high-risk environments. Once thought to be "noise" in the cybernetic communication system as described by Shannon and Weaver, workers themselves can be considered expert at recognizing, if not articulating, danger. Sauer devotes an entire chapter to describing hand and body gestures that indicate a rich site for further exploration of embodied knowledge and of attempts to articulate worker knowledge as expert knowledge. The challenge, as Sauer describes it, comes as much from the resistance of powerful managers who would have to rethink their own relative expertise and understanding of the mine-as-economic-engine as it does with the miners' perspectives of the mine-as-workplace, and indeed, the underground tunnels as miners' lifeworld.

Sauer's book makes an important contribution to this chapter's argument, first by recognizing the need to include more and different sources of information in the design of risk-avoidance documentation. Second, the book offers strategies to begin creating (inventing) means of incorporating information from these various newly recognized sources into authoritative discourse. These sources, commonly thought of as noise, or even of interference, are generally marginalized: as labor, a cost in the economic charting of the workplace, or as interference, overly cautious regulation of the workplace. Workers are generally discounted as possible sources of insight. Rather than viewed as noise to the system, workers are brought in and represented as important, valuable, indeed necessary to the system, for without them the system utterly fails. Risk communication becomes a collaborative process between expert mine workers and expert mine managers, with communication experts mediating their working communications and environments. Of course the opposite—knowledgeable managers passing crucial information down to ignorant workers—is a caricature of the relationship, but unfortunately the representation is not far off the conditions of some mines that Sauer describes.

Documentation, then, is not limited to getting information down the institutional hierarchy but can also flow up the hierarchy, a two-way process that is crucial in the attempt to prevent unknown dangers. The hierarchy still exists, especially in terms of pay and benefits, but the flow of information between strata of employees, management and labor, is no longer hindered; instead, it is enabled by the work of the technical communicator.

Readers may be tempted here to represent the bridging work done by rhetoricians as translation between different constituent parts of the organization, or even as the creation of a communication conduit between what have previously been separate entities, unable to speak. However, these metaphors for the technical communicator's work have been effectively deconstructed elsewhere (Slack, Miller, and Doak; Johnson-Eilola). Rather, the technical communicator takes raw data, whether in the form of worker impressions or even of gestures, and transforms that data into usable knowledge. The tools of postmodern digital communication are merely means of achieving this communication, the just-in-time informating of the organization. While I would not advocate that action be limited to functional transfer of information, it is an important role that needs to be played, particularly in the short term. See Robert Johnson's description of three long-term responsibilities for technical rhetors, which he names as redefining growth, becoming stewards of technology, and becoming responsible to broader cultural contexts. But before we can steward technology and culture, we have to effectively and competently communicate about existing technology, its risks, and its contexts.

Finally, Sauer represents technical communication at the point of invention, which, as one of the five classical canons, directly invokes rhetoric. The challenge facing technical rhetors is not only to document known dangers and to prevent disaster from reoccurring, but to create mechanisms for forecasting danger and thereby avoid the loss of capital, the damaging or maiming of human beings in accidents, and the loss of life. Forecasting can never be certain, but we can use rhetoric to focus on the *probable* outcome among many *possible* outcomes of danger rather than simply recording dangers that are known and preventing reoccurrences of past tragedies. This is no easy task and recenters the concerns of the field not just on tools or theory (or even pedagogy), but on a mode of inquiry that first analyzes existing cultural situations, processes these findings, and then designs future configurations in order to avoid the problems articulated in the original analysis. It is cultural studies, represented here as a methodology for articulating and critiquing existing relations among people, technologies, and institutions, wed to design, the future-oriented activity through which new relationships are made.

AUDIENCE-CENTERED DESIGN: THE DEMISE OF THE ARTISTIC GENIUS

For design, the problem of postmodernity remains one of authority. Designing something as complex as a system, in an age of emergent systems and distributed authority, often involves examining how systems (be they social, technological, institutional, etc.) develop on their own without expert intervention. But how can one claim to know what is best, or what solution will work, in any one situation? And how can one ensure against doing no harm to the human beings who make up the population that will be most directly affected by the system designed and put in place by experts?

This is where the shift is occurring in the definition of expertise. In democratic culture, all citizens are equal under the law, but some are set apart by knowledge, experience, and training; some offer informed, validated advice to populations; and some offer solutions appropriate for specific situations. These needs are revealed as sites for intervention, and I have described the process for locating such power/discourse/action bottlenecks in previous research (Salvo 2004). This chapters calls on two different but parallel discourses, design and cultural studies, to come together to form a dialogic and potentially democratic justification for rhetorical action. In so coming together, cultural studies research and pedagogy are shifted from analytic methodologies to components of critical rhetorical action. Kaufer and Butler have described this shift as reorienting rhetoric, in their terminology, as an art of design. Kaufer and Butler assert that the technical "rhetor is the architect of the social world" (74).

McQuail is referenced earlier in this chapter inviting cultural studies researchers to participate in policymaking. The cultural studies policymaker McQuail envisions is akin to the rhetor described in *Rhetoric and the Arts of Design*, yet policymaking does not go far enough. According to Kafer and Butler, time becomes an important element in designing the social world:

> Aristotle spoke of different kinds of rhetoric as focusing on the past (judicial), present (encomium), and future (policy) respectively. But in constructing the social world, the rhetor must design the here and now to accommodate every horizon of time so as to influence the audience's decision-making. The rhetor must be

> timeless enough to retrofit the audience to its historic past and destined future, timely enough to foil any live opponent with similar but competing goals, and well enough timed to give the audience pleasure in the now. (Kaufer and Butler 74)

Analysis of the past, as in Kynell-Hunt and Savage's collection, coupled with the analysis of the present, as presented in Henry's *Writing Workplace Culture*, with an eye toward the future, as in Sauer's *Rhetoric of Risk*, allows a critical technical rhetor, such as the one described in Faber's *Community Action*, to design for the future. Kaufer and Butler's designing rhetorician is an impossibly competent expert, effectively building on the knowable past, in the understood present, and for the foreseeable future. But as a goal for rhetorical practice, the timeliness and powerfulness of the rhetor's vision is nevertheless compelling.

Postmodernism's uneasy relationship with authoritative discourse complicates this move in time, however. What basis does rhetoric have to act in the name of the populations who will be knitted up in the systems of discourse/technology/culture? Even if we concede to view rhetoric as an art of design, the postmodern question remains: What distinguishes the discourses of rhetoric from other discourses that seek power over people? However well intentioned, many discourses of design tend toward designing worlds *for* people to inhabit rather than designing worlds *with* the people who will be inhabiting them.

Accepting design-as-world-building as a part of cultural studies discourse fails unless accompanied by a shift in authority from designing *for* consumption to designing *with* stakeholders. Rather than be stymied by postmodern paralysis, the argument develops an "amodern" position parallel to that developed by Bruno Latour in *We Have Never Been Modern*, one that repositions the communication expert *among* her clients rather than, as with modern discourses, above or beyond the populations that professionals are supposed to serve. Such a reconfiguration of the rhetorician-as-designer positions him or her as one among many, one among equals.

CONCLUSION: A TRANSACTIONAL RHETORIC FOR CULTURAL STUDIES

I opened this chapter with a reference to Berlin's definition of postmodern literacy without tracing how or why Berlin's definition might be considered postmodern or, indeed, why readers should be concerned about whether we are or are not in an age of postmodernity. Latour argues that, having never been modern, it is simply wrong to talk about being beyond, and so *post*modernity is simply a misnomer. But the peril of postmodern illiteracy that Berlin points to is real, whether or not we accept the accuracy of its placement in postmodern circumstance. Though meant to enable democratic participation, Berlin's call to action in response to this illiteracy nevertheless contradicts its own postmodern placement: How can one act? How does one muster the authority, the justification to act, when the notion of authority itself is being questioned? I am not the first to pinpoint this problem in the postmodern, but it is important and valuable to rehearse this argument here because it has remained a defining sense of what the postmodern is: the source of *intellectual* paralysis. Zygmunt Bauman captures this danger when he states that, "In the end, a *universal dismantling of power-supported structures has been the result.* No new and improved order has emerged, however, from beneath the debris of the old and unwanted one" (viii–ix).

In the ruins, whether the general ruins of postcapitalist culture or the ruins of the traditional university, after deconstruction has cleared the intellectual moorings of modernism and the justification of its expression of power, nothing is left that withstands the force of postmodern critique. At least, according to Bauman, intellectuals have been unable to justify anything to each other, and this distinction between the postmodern as an intellectual state of mind and as a cultural reality continues to be Bauman's contribution to the conversation. It is not that the world has stopped and decided to wait for philosophers and cultural critics to build something new and worthy of belief on which to structure a new postmodern culture. Indeed, it is most disconcerting to intellectuals that the rest of culture moves on, fairly unaffected and certainly uninterested in this postmodern crisis of authority.

But it is not sufficient to accept paralysis, to accept the loss of authorizing narratives, under any name. Instead, it is important to look at what may be left on the ground and in the ruins, some sprout or twig from which new life and new structures can be coaxed, in part to reinvigorate the role of the academic to move beyond dispassionate observer and reimagine intellectuals *as* citizens.

And so I am looking to technical and professional communication to develop a practical rhetoric that responds to the failures of ludic postmodern discourse, to try to offer, if not a "new and improved" claim to authority, then a postmodern authority that allows both critique and action, observation and participation, deconstruction alongside construction. Postmodern criticism points to the markers that would allow an authoritative discourse to remain. What would it take for action to be ethical in a postmodern world? Again, Bauman offers a hint in his attention to what he terms the "ethical paradox of postmodernity":

> The ethical paradox of the postmodern condition is that it restores to agents the fullness of moral choice and responsibility while

> simultaneously depriving them of the comfort of the universal guidance that modern self-confidence once promised. Ethical tasks of individuals grow while the socially produced resources to fulfill them shrink. Moral responsibility comes together with the loneliness of moral choice. (xxii)

Bauman asserts that, even as the ideas on which the measurement of "right" and "wrong" are based are being dismantled, postmodernity is unsettling because we are expected to act ethically and responsibly. How do we know if we are acting responsibly? How do we know what *ethical* means? And yet we must bear the consequences of our actions, as Sullivan and Porter so effectively describe in *Opening Spaces*, which calls for revealing (to the extent possible) the researcher's own placement and assumptions in working with particular materials in a certain way. This involves revealing not only how research and design has been performed in a certain way, but also discussing why decisions have been reached: in other words, postmodern action is always striving for openness, so that citizens decide for themselves whether action is legitimate.

One response to the postmodern decentering of authority is nihilism—nothing matters, and there can be no consequences for action, good or bad, because there are no reliable measures for these categories. Another response is radical subjectivity—everyone can do what he or she wishes because there are no ways to measure whether this or that action is better or worse; there is simply change. Jürgen Habermas has articulated another position, one that is conservative in saying that modernism is a painstaking accomplishment of culture that cannot be so easily deconstructed, or at least should not be so easily dismissed, and ruins or no, rationality is all we have. These modes of thinking still offer no way out of the paralysis of the postmodern legitimation crisis.

And so the question remains: On what do we base action and authority? What foundation allows enough authority and enough justification for action? If Bauman is right, and it is Auschwitz and Berkinow that complicate postmodern notions for authority rather than Jameson's "cultural logic of late capitalism," in which economic realities have overtaken cultural commentators, then postmodern paralysis is a chimera, a habit of mind. Any world building exercise—that is, any attempt to design the world to be inhabited by others—carries the ethical load of (in Adorno's weighted construction) avoiding another Shoah. It is thinking that the possibility of fascism necessitates arriving back at fascism that debilitates action. Bauman views the debilitation as a habit of mind rather than the actual probability that fascism will arise from world-building. Ethics, for Bauman, makes action possible, and in sympathy with Adorno one may assert that the prime ethical requirement is that

Auschwitz never happen again. One way of fighting totalization and silencing of the weak is to foreground a rhetoric of dialogue in which all stakeholders are encouraged to raise their voices, to engage in the cultural as well as the technical aspects of technology, with the expectation that technological participation will support wider cultural participation.

Jameson and Bauman define two perspectives in an ongoing dialogue regarding the status of the postmodern in postindustrial culture that illustrates the distinctions I am drawing between active engagement and reactive commentary. Engagement requires risk and requires the rhetor to stake a claim and propose an outcome different from the extant reality. Here, in accepting optimism for change, is where rhetoric as an art of design breaks with rhetoric as an act of postmodern despair, and where hopefulness of progressive ideas contrasts with the hopeless conservatism of Habermas's desperate clinging to shards of rationality. Rather than despairing at the worst possible outcomes that we feel inexorably damned to inherent, we can instead build possible futures, *design* possible futures that we would like to inhabit. What possible human-scaled worlds are we likely to bring to fruition through cooperation with citizens like ourselves? In accepting that others share our hopes and aspirations, we can participate in the larger discourses that will indeed build our future cities, workplaces, and communities.

At the start of this chapter, I offered a rewritten and updated version of Berlin's definition of critical literacy:

> Learning to gain some control over communication forms and the technologies that enable them, students become active agents of social, political and technical change, learning that social and technological worlds have been made and can thus be remade to serve the interests of democratic society.

This definition does not privilege one vision of the best world to which others must accommodate themselves, and so in that sense is not utopian but responsive to contradictory visions. In "Writing Classroom as A&P Parking Lot," Geoff Sirc offers a postmodern definition of literacy that is urban in its diversity of perspectives and that respects students' idiosyncratic visions of what should and ought to be. In technorhetorical context, rearticulated expertise requires respect for diversity and for communities' self-determination and self-articulation. We can help articulate how best to realize the goals communities have set for themselves rather than hijacking their vision, being alert to but not paralyzed by the ethical limits of design, to participate in the world as it is being remade while remaining conscious of what may go wrong, and how world-building can be corrupted and co-opted by empire-building. Inaction is as dangerous as wrong action, and observing and recording change is no role for rhetoric.

Cultural studies offers technorhetoricians a powerful analytic methodology. However, to remain rhetoric, action must be added to analysis. Rhetoric, as I have described it here, engages in the rearrangement of people, technologies, and institutions through discourse to better accomplish civic goals (in addition to those of commerce). And in taking such action, technorhetoricians can take elements of cultural studies methodology in order to better inform our designs of rhetorical engagement, while our expertise in the workplace offers cultural studies a much-needed opportunity to engage the world beyond either the criticism of texts enjoyed in times of leisure or passive semiotic criticism of technological and workplace texts. And in so offering, we become better citizens both by expanding our sense of possibility for clients with whom we work as expert advisers and for the academic world with which we work as knowledge producers and explorers of the limits of epistemology.

NOTE

1. Bruno Latour and Donna Haraway offer provocative exceptions to this in *Aramis, or the Love of Technology* and *Simians, Cyborgs, and Women*, respectively. In articulating the constellation of individuals and institutions (and their discourses) responsible for the demise of the French transit system, Latour points to the need for rhetorical intervention that would help future workers, planners, and thinkers avoid such a demise. In the latter, especially in "Situated Knowledges," Haraway rearticulates scientific objectivity by reinterpreting the tools and context of scientific inquiry as objects and tracing the discursive constructions that make the laboratory a privileged site of inquiry.

WORKS CITED

Adorno, Theodore. "Education After Auschwitz." In *Critical Models: Interventions and Catchwords.* Trans. Henry W. Pickford. New York: Columbia UP, 1999.

Althusser, Louis. "Ideology and Ideological State Apparatus." *"Lenin and Philosophy" and Other Essays.* Ed. Louis Althusser. London: New Left Books, 1977.

Barry, Ann Marie Seward. *Visual Intelligence: Perception, Image, and Manipulation in Visual Communication.* Albany, NY: SUNY P, 1998.

Berlin, James A. *Rhetorics, Poetics and Cultures: Reconfiguring English Studies.* Reprint (Lauer Series in Rhetoric and Composition). West Lafayette, IN: Parlor Press, 2003.

Bauman, Zygmunt. *Intimations of Postmodernity.* New York: Routledge, 1992.

Carey, James W. "Reflections on the Project of (American) Cultural Studies." *Cultural Studies in Question.* Ed. Marjorie Ferguson and Peter Golding. Thousand Oaks, CA: SAGE, 1997.

Cyphert, Dale. "Excavating the Writerly Workplace." Review of Jim Henry's *Writing Workplace Cultures: An Archaeology of Professional Writing. Review of Communication* 2.1 (2002): 26–30. http://www.natcom.org/pubs/ROC/one-one/Vol2No1.htm.

Faber, Brenton D. *Community Action and Organizational Change: Image, Narrative, Identity.* Carbondale, IL: Southern Illinois UP, 2002.

Feenberg, Andrew. *Transforming Technology: A Critical Theory Revisited.* 2nd ed. of *Critical Theory of Technology*. New York: Oxford UP, 2002.

Ferguson, Marjorie, and Peter Golding, eds. *Cultural Studies in Question.* Thousand Oaks, CA: SAGE, 1997.

Garnham, Nicholas. "Political Economy and the Practice of Cultural Studies." In *Cultural Studies in Question.* Eds. Marjorie Ferguson and Peter Golding. Thousand Oaks, CA: SAGE, 1997.

Habermas, Jürgen. *Legitimation Crisis.* Trans. Thomas McCarthy. Boston: Beacon P, 1990.

Hall, Stuart. "Encoding/decoding." *Culture, Media, Language: Working Papers in Cultural Studies, 1972–79.* Centre for Contemporary Cultural Studies. London: Hutchinson, 1980. 128–38.

Haraway, Donna J. *Simians, Cyborgs, and Women: The Reinvention of Nature.* New York: Routledge, 1991.

Henry, Jim. *Writing Workplace Cultures: An Archaeology of Professional Writing.* Carbondale: Southern Illinois UP, 2000.

Herndl, Carl G. "Introduction: The Legacy of Critique and the Promise of Practice." *Journal of Business and Technical Communication* 18.1 (2004): 3–8.

Jacobson, Robert. *Information Design.* Cambridge, MA: MIT P, 1999.

Jameson, Fredric. *Postmodernism, or the Cultural Logic of Late Capitalism.* Durham, NC: Duke UP, 1992.

Johnson, Robert R. *User-centered Technology: A Rhetorical Theory for Computers and Other Mundane Artifacts.* Albany, NY: SUNY P, 1998.

Johnson-Eilola, Johndan. *Nostalgic Angels: Rearticulating Hypertext Writing.* Norwood, NJ: Ablex, 1997.

Kaufer, David, and Brian S. Butler. *Rhetoric and the Arts of Design.* Mahwah, NJ: Lawrence Erlbaum, 1996.

Kynell-Hunt, Teresa, and Gerald J. Savage. *Power and Authority in Technical Communication Volume I: The Historical and Contemporary Struggle for Professional Status.* Amityville, NY: Baywood, 2003.

Latour, Bruno. *Aramis, or the Love of Technology.* Cambridge, MA: Harvard UP, 1996.

———. *We Have Never Been Modern*. Cambridge, MA: Harvard UP, 1993.

McQuail, Denis. "Policy Help Wanted: Willing and Able Media Culturalists Please Apply." *Cultural Studies in Question.* Ed. Marjorie Ferguson and Peter Golding. Thousand Oaks, CA: SAGE, 1997.

Olsen, Gary A. "Interview with Clifford Geertz." *JAC* 11.2 (1991): http://jac.gsu.edu/jac/11.2/Articles/geertz.htm.

Salvo, Michael J. "Ethics of Engagement: User-Centered Design and Rhetorical Methodology." *Technical Communication Quarterly* 10.3 (2001): 273–290.

———. "Rhetorical Action in Professional Space: Information Architecture as Critical Practice." *Journal of Business and Technical Communication* 18.1 (2004): 39–66.

Sauer, Beverly. *The Rhetoric of Risk: Technical Documentation in Hazardous Environments.* Mahwah, NJ: Lawrence Erlbaum, 2002.

Schwartz, John. "More Lawsuits Filed in Effort to Thwart File Sharing." *New York Times* Late Edition. 24 March 2004: C4.

Shannon, Claude E., and Warren Weaver. *A Mathematical Model of Communication.* Urbana, IL: U of Illinois P, 1949.

Slack, J. D., Miller, D. J., Doak, J. "The Technical Communicator as Author: Meaning, Power, Authority." *Journal of Business and Technical Communication* 7.1 (1993): 12–36.

Ulmer, Gregory L. "Foreword/Forward (Into Electracy)." *Literacy Theory in the Age of the Internet.* Ed. Todd Taylor and Irene Ward. New York: Columbia UP, 1998.

Wurman, Richard Saul. *The Information Architects.* New York: Watson-Guptill, 1996.

Chapter Ten

Extending Service-Learning's Critical Reflection and Action

Contributions of Cultural Studies

J. Blake Scott

Service-Learning's Potential

Service-learning approaches to technical communication continue to proliferate, as demonstrated by the textbook *Service-Learning in Technical and Professional Communication*, the special issue of the service-learning journal *Reflections* that focuses on our field, and other recent fruits of this application. Since the earlier calls by Thomas Huckin and Leigh Henson and Kristene Sutliff, more and more teachers are discovering that their technical communication courses—with emphases on workplace writing, collaboration, and project management—are particularly fertile sites for service-learning.

According to the National and Community Service Trust Act of 1993, service-learning is "a method by which students learn through active participation in thoughtfully organized service; is conducted in, and meets the needs of the community; is integrated into and enhances the academic curriculum; includes structured time for reflection and helps foster civic responsibility" (859–860). Although the relationship between writing and service can take multiple forms, most technical communication teachers have students mainly write *as* their service rather than just about it. One of the key components of service-learning is active, contextualized learning, something also valued in technical communication pedagogy. For an increasing number of teachers, service-learning seems like a natural extension of sociorhetorical curricula involving cases and real-world writing assignments. As Thomas Deans points out, service-learning presents students with additional, community-based audiences, discourse communities, rhetorical contexts, modes of collaboration, and ethical issues (9–10).

Although both service-learning and technical communication have been influenced by a pragmatism that values experiential learning and social action, this pragmatism has taken very different forms in the two pedagogical traditions. As my co-editors and I argue in the introduction to this collection, technical communication pedagogy has been driven largely by a hyperpragmatism that values conformity and rhetorical utility and that focuses on students' successful enculturation into workplace discourse communities. In addition to squelching ideological and ethical critique, hyperpragmatism can narrowly position students as preprofessionals and limit the focus of the course to the immediate rhetorical situations and production tasks at hand.

A small but growing number of technical communication scholar-teachers have turned to the tradition of service-learning to develop less utilitarian and hyperpragmatic pedagogical approaches (see Dubinsky, Scott, Kimme Hea). Indeed, service-learning was developed in part as a reaction to the overvocationalism of higher education (Liu), though its institutionalization sometimes threatens to connect it to internships and co-ops at some universities. Although service-learning values praxis, or social action, it also entails much more than this, as the definition above indicates. Among other things, this praxis must meet a community need and must be accompanied by structured reflection about civic responsibility. We might say that, along with praxis, service-learning is concerned with the development of phronesis or ethical, wise judgment. Thus, service-learning can expand the vocational focus of technical communication to a more broadly civic one, and it can enhance praxis with critical reflection. Deans traces service-learning's pragmatist heritage to John Dewey, who sought to integrate active experimentation and reflective thought (31). In addition, his notion of social action was inherently civic, political, and reform-minded. For Dewey, the ultimate goal of education was not individual professional success but civic involvement and democratic social reform. Like Cicero, Dewey thought education should train students to be effective and *ethical* citizens engaged in the life of the polis. Given this lineage, service-learning may have more in common with cultural studies, especially the Freirian tradition of radical pedagogy, than with technical communication. Cultural critics Lawrence Grossberg and James Carey describe Dewey as an important early figure in American cultural studies, pioneering work on relations between communication and social life (Grossberg 144; Carey 64).

Several technical communication scholars have celebrated the benefits of service-learning, mostly pointing to the ways it enhances such pragmatic skills as audience awareness, interpersonal communication, and project management. Huckin, for example, explains that service-learning provided his students with rich opportunities to learn about interpersonal communication and organizational politics and to develop more rhetorically sensitive writing strategies. Catherine Matthews and Beverly Zimmerman report from their qualitative study that service-learning improved their students' academic learning, especially their rhetorical skills. In our textbook, Melody Bowdon and I similarly advocate service-learning as a way to enhance students' rhetorical, collaboration, and project management skills.

Less common are sustained illustrations and explanations of how service-learning teaches students civic awareness and ethical civic action. Most scholars touting service-learning's benefits describe its civic dimension as a motivator. Huckin notes that his students were motivated by the "civic pride and satisfaction they felt from helping their fellow citizens" (54). Matthews and Zimmerman emphasize the benefit of student motivation even more, arguing that students take a more active role in their learning. Henson and Sutliff suggest that service-learning can even motivate students beyond the course to "ultimately develop reasonable goals for their own roles in improving the world in which they live" (192). When framed in terms of student motivation and self-discovery, however, the civic aspects of service-learning can be muted, as I will later explain.

These celebratory accounts of service-learning notwithstanding, the transformative potential of service-learning, and especially the component of reflection, goes unfulfilled in many technical communication courses, including my own. Experienced service-learning teacher Jim Dubinsky describes the previous versions of his own professional communication course as an example of the widespread failure of such courses to engage students in critical reflection about citizenship, "in being responsive to the community needs" (69). As Dubinsky's description illustrates, service-learning approaches can all too easily lose their critical, activist edge and civic scope. In some cases these approaches can be coopted by the very hyperpragmatism they seek to challenge. In other cases they can be limited by liberal foci on personal development and discovery that leads to shallow reflection. After discussing these limitations in more detail, I'll argue that cultural studies theory and heuristics can offer valuable remedies. Integrating cultural studies into a service-learning approach can go a long way toward addressing the "problem" of reflection and can lead to a more robust critical pedagogy of civic engagement and sustained ethical intervention. As Donna Bickford and Nedra Reynolds put it, we need to base service-learning on "models of citizenship that combine critical consciousness with action and reflection" (2).

Service-Learning's Unfulfilled Promise

Despite service-learning's partly critical heritage, it is often co-opted by a hyperpragmatism that moves past ethical deliberation about the larger cultural conditions, circulation, and effects of technical communication. The logistical demands of a service-learning course, students' desire for a practical education, the disciplinary emphasis on uncritical accommodation, and the institutional and cultural emphasis on preparation for corporate success all work to maintain a pedagogy that facilitates praxis but not phronesis.

As those of us who teach service-learning know well, the practical demands of service-learning projects can be challenging, especially when such projects are carried out within the span of a semester. The teacher and students must work with a sponsoring agency or organization to negotiate a project, develop it, produce it, manage it, and evaluate it. Students must analyze their intended audience and the discourse community of their sponsoring organization, studying its values, discourse conventions, and writing processes. Often, students must learn how to produce new genres and use new media. They must also enact complex collaboration and project management strategies as they balance the expectations of their course and sponsoring organization. In the version of service-learning endorsed by Huckin, Bowdon, and me, as well as others, students not only produce communication projects for their sponsors, but they also produce a proposal, progress report, evaluation report, and possibly other documents as part of their project invention and management. Teachers of service-learning face the challenges of coordinating and facilitating projects, maintaining relations with community sponsors, teaching students about various genres and conventions, and evaluating student work. In the face of such demands (in place to ensure students' pragmatic success), it's easy to see how service-learning courses can leave little time for critical reflection.

In treating the service-learning experience as "practice" in learning to write for an organization, many service-learning technical communication courses encourage students to focus on successfully negotiating their immediate rhetorical situation and the new discourse community of their sponsoring organization. Accommodating the expectations of the organization (and the course) is often students' foremost concern. This can lead to a narrow understanding of "accommodation"—a notion central to much technical communication pedagogy—as fulfilling set duties, goals, and conventions rather than engaging, responding to, and empowering the stakeholders of one's work. Sometimes students are so eager to work with and please their community sponsors that they overlook the ethical implications of their work for other stakeholders, including their texts' primary users. In this way, students are positioned as citizens of the organization but not of the larger community.

One way hyperpragmatism can coopt service-learning's potential is to subsume reflection—one of the keys to civic awareness—into uncritical invention and project management exercises. Students might reflect on how their sponsoring organization's ethos and values shape the parameters of their rhetorical choices. They might reflect on their attempts to understand and conform to the organization's discourse conventions. Or they might analyze the interpersonal relationships at their organization as they plan their collaborative writing processes. Such activities can indeed be useful, but they do not necessarily prompt ethical critique and intervention. To aid their project management, students might be asked to reflect on their collaboration with each other and their agency contacts, for example, or on specific challenges they're facing and how they might respond to these challenges.

Because this type of reflection focuses on the pragmatics of the production process, it doesn't address the larger cultural forces shaping students' texts and the subsequent transformations and effects of these texts as they are distributed and taken up. We need strategies that not only infuse critique into reflection but also help extend the scope and trajectory of this critique. This extension is necessary for ensuring the ethical sustainability of students' service-learning projects.

Even when reflection is not subsumed into the pragmatics of the course, it can still be rendered a shallow, uncritical exercise. Some courses just have students keep personal discovery journals or, worse, logs that simply document their experiences. Often these courses present service as charity work and a mechanism for personal growth; indeed, versions of reflection that focus on personal growth correspond nicely to technical communication's liberal emphasis on student development. For example, students might reflect on what they've learned about themselves and their community in a limited and often self-congratulatory way, answering such questions as "How has this experienced changed you?" and "What difference have you made to the organization and community?"

Bruce Herzberg and others have critiqued models of service-learning that focus on students' personal development and that encourage students to adopt empathic and charitable points of view. Bowdon and I warn students to beware of "the seduction of empathy" (her term), which can cause them to overestimate their understanding of community problems and prevent them from engaging the unique contexts of those with whom they work (5). Aaron Schultz and Ann Ruggles Gere similarly warn both teachers and students of seeing themselves as "liberal saviors" (133).

In addition to giving students the misconception that they can march in and solve community problems, a philanthropic or charitable viewpoint can cause students to ignore "the structural reasons to help others," explain Bickford and Reynolds (230). Indeed, Bickford and Reynolds argue that many students are uncomfortable viewing their work as activism and unable to see their work as a "contribution to structural social change" (238). Conditioned to view their education through a hyperpragmatist lens, technical communication students can be even more resistant to the role of community activist, instead preferring to view themselves as "consultants" and their sponsoring organization as "clients." Dubinsky explains that, before he infused his service-learning course with an emphasis on civic virtue, his students "acted more like consultants for hire. They talked about *working for* clients rather than *working with* partners" (69). Beyond limiting students' civic engagement, this preference can manifest in students bestowing their "expertise" on the organization rather than recognizing their lack thereof. Especially susceptible to these problems are "hit and run" service-learning projects that don't encourage students to engage community stakeholders throughout the process or to ensure the long term viability and responsiveness of their work. In addition, versions of reflection that focus on personal growth correspond nicely to technical communication's liberal emphasis on student development.

Some advocates of service-learning in technical communication are more diligent about positioning students as civic workers and pushing them to deliberate more deeply and critically about the ethics of their work, its exigency, and its effects. Huckin, for example, has his students discuss the larger problems addressed by their work, asking them to interrogate the structural causes of these problems and the need for their organization. Henson and Sutliff similarly ask students, "What social conditions have created the need for the nonprofit you served?" In their argument for addressing social structures, Bickford and Reynolds advocate supplementing the question, "How can we help these people?" with the more complex question, "Why are conditions this way?" Such questions certainly point to more critical, robust reflection, but they are often deemphasized in service-learning courses for being more tertiary than integral. Such reflection often takes place toward the end of the project or course, when it is too late to act on. Huckin advises teachers to pose his questions in project wrap-up discussions, Henson and Sutliff present their questions as prompts for an "exit report," and Louise Rehling has her service-learning interns reflect on their projects in final essays.

CULTURAL STUDIES ENHANCEMENTS

In my own teaching, I have increasingly turned to cultural studies theory, methods, and heuristics to ensure the critical and civic potency of service-learning. Like service-learning, cultural studies is based on multiple traditions and can take various forms. Some of these forms bear little resemblance to service-learning and its emphasis on rhetorical action. The cultural studies pedagogy of James Berlin, for example, has been critiqued by Susan Miller and others for positioning student as pseudo-critics rather than civic rhetors and for privileging semiotic analysis over rhetorical production. In his argument for instilling in technical communication students an "aptitude for cultural criticism," Jim Henry critiques some forms of cultural studies for their "insistence on forming students as critical discursive consumers all the while wholly ignoring their formation as critical discursive producers in any genre other than the academic essay" (215 this collection).

Not all cultural studies approaches deemphasize public action. Cultural critic Stuart Hall, for example, laments the tendency in American cultural studies to overrely on semiotic analysis ("On" 149). In contrast to this tendency, many cultural studies traditions, from the culturalism of the Birmingham school to the poststructuralism of Foucault, view political action as integral to their methodology. In this way, rhetoric and cultural studies are not as different as many in our field assume.

The cultural studies model from which I work shares with service-learning a commitment to both critical awareness and ethical action. Lawrence Grossberg offers one of the most succinctly useful definitions of cultural studies, in my view, and one that corresponds to my approach. Cultural studies, he explains, is concerned with the "ways 'texts' and 'discourses' are produced within, inserted into, and operate in the everyday lives of human beings and social formations, so as to reproduce, struggle against, and perhaps transform the existing structures of power" (237).

To elaborate on this definition, cultural studies accounts for the cultural circulation of texts. This part of Grossberg's definition invokes cultural critic Richard Johnson's notion of the cultural circuit, which tracks the trajectory of cultural forms as they are produced, distributed, consumed, and integrated into people's lives, where they become part of shifting conditions for production. Cultural forms are rearticulated or transformed as they circulate, taking on new functions and enabling and disabling new effects. The cultural circuit, I have argued elsewhere, can help students be more attentive and responsive to technical communication's shifting conditions, instantiations, functions, and effects beyond the scope of the project or course ("Tracking").

Another important, related element of the definition above is its emphasis on critiquing structures of power. This can involve critiquing the ways organizational, institutional, or larger social structures regulate technical communication and shape its effects. James Porter et al. argue for critique that similarly focuses on "particular institutional [power] formations that are a local manifestation of more general social relations" (621).

Further, a cultural studies critique pays particular attention to how power shapes identities and relationships. For example, Henry's students critique how organizational structures and processes shape the identities and work life of technical communicators, often in marginalizing ways. James Porter et al. and Bickford and Reynolds call on us to engage in spatial and boundary analysis, in both the material and ideological sense. The former ask how boundaries are "constructed, maintained, expanded, and challenged" by power structures, and the latter ask how configurations of space reflect and create hierarchical relationships and power differentials. My students often critique the hierarchical relationships created by institutional *processes* as well as spaces.

Importantly, Grossberg's definition of cultural studies also emphasizes a call to action. Porter et al. argue that institutional critique must include an "action plan" for transforming problematic policies (613). Even before formulating such a plan, however, students in a service-learning course should begin their projects with the ethical imperative to transform as well as critique. In the vein of Patricia Sullivan and Jim Porter, I encourage my students to consider the ethical imperatives of respecting difference, being attentive and responsive to others, creating more egalitarian and just practices, and empowering those affected by their work (110).

This elaboration of a cultural studies approach points to several ways integrating cultural studies and service-learning can resist the pull of hyperpragmatism and liberal shallowness. First, cultural studies can help expand students' invention to include deliberation about the broader cultural conditions and circulation of their work. In addition to focusing on their immediate rhetorical situations and production process, students could track the shifting functions of their texts as they are transformed by various actors. Beyond framing their projects as solutions to the communication needs of their sponsoring organizations, for example, students could frame them as responses to broader social exigencies. Along with the instructor, students can interrogate the institutional and larger cultural conditions—including capitalist and countercapitalist values and forces—that have contributed to the development of service-learning at their university and beyond.

Second, cultural studies can help ensure that students' reflection is critical and civic-minded rather than mainly pragmatic or expressive. Guided by what Michael Salvo calls an "ethic of engagement," students could assess, for example, how their production process excludes some perspectives and privilege others. They could assess how their texts might position and affect the practices of various users and to what effects. Although students' critique would largely focus on their own work, they would situate this work in larger institutional and social power relations. Thus their reflection could consider the larger social structures that create and maintain community problems, as Bickford and Reynolds suggest, or the institutional policies that constrain their responses to these problems in particular ways.

Finally, cultural studies can help push students to translate their critiques into ethical rhetorical action throughout their projects, revising as they go along. Students could develop strategies for involving other stakeholders in the planning and assessment of their projects, in particular. Such action could be informed by the method of intercultural inquiry developed by Linda Flower and others at Pittsburgh's Community Literacy Center. Intercultural inquiry involves inviting stakeholders in a community problem to be partners in a problem-solving dialogue that values their situated knowledges (Flower). Ideally, this dialogue would begin early in the service-learning project. Even when their project constraints make implementing such a dialogue difficult, students could recommend ways for the organization to do so in future projects. Through their emphasis on rhetorical action and its real-world contexts and effects, technical communication projects, and especially service-learning versions, can serve as powerful counterexamples to cultural studies work that remains safely in the realm of academic critique.

Bickford and Reynolds pose what I think should be a central question for students in a service-learning course: How can we enter relationships "in ways that help destabilize hierarchal relations and encourage the formation of more egalitarian structures?" (241). This question calls students to do more than fulfill the expectations of the organization and instructor or even accommodate their targeted audience; it also calls them to be community *advocates* who work with other stakeholders for structural reform on organizational, institutional, and/or social levels. Such reform is no small task, of course, and involves more risk—of overwhelming students, of creating conflict among stakeholders, even of exacerbating student hubris—than the typical pragmatic approach. First engaging in a problem-solving dialogue can help students approach their advocacy humbly, recognizing that they cannot simply march in and solve the problem.

Pedagogical Applications

The turn to cultural studies in my service-learning course has led me to revise several of my staple assignments and incorporate new ones. My courses typically follow the model outlined in my textbook with Bowdon: small groups of students engage in semester-long technical communication projects (broadly defined to include marketing and business communication as well as various media) for nonprofit community and campus organizations. Over the course of the semester, students in my service-learning courses also engage in various reflection activities and complete a series of project management assignments that usually include a proposal, discourse analysis, progress report, and evaluation report.[1] In past courses, these assignments were mostly pragmatic in nature, though some did require students to deliberate about any ethical challenges they faced.

When I first began incorporating cultural studies concerns and heuristics, I mainly tacked them on as extra reflection activities at the end of the semester. For the most part, I relegated reflection to a final evaluation report that encouraged students to critique the ethics of their process and texts but did not invite them to do take or recommend any corrective action. If this evaluation report made students critically self-reflexive, this reflexivity was an afterthought to the "real work" of their projects and more of an exercise they completed for their instructor and grade.

In more recent courses I've tried to more fully integrate cultural studies concerns into major assignments and the very processes that students engage in over the course of their projects. I've found that project management and invention assignments, when revised to move beyond narrow pragmatic concerns, can offer students compelling exigencies for cultural critique and forward-looking action.

One of the first major assignments in many service-learning technical communication courses is a project proposal. The main purpose of this assignment is for students to develop, in conjunction with their instructor and sponsoring organization, the "what" and "how" of their projects—what they will produce and how they will do it. Because most service-learning projects are difficult to complete within a semester even when students begin right away, many instructors work with organizations to determine the parameters of the "what" ahead of time. Students still need to show that they understand the instructor's and organization's expectations and to determine some details about their "solutions," but they may not be asked to explore various alternative solutions or the larger community problems that gave rise to the need in the first place (the "why," if you will). All too often, they are rushed through the proposal writing process. In addition, students are often told that their proposal must function as a blueprint for their project and even a contract with their instructor and organization; as a result, students are likely to see the proposal as a fixed document.

Reframed through a cultural studies lens, the proposal assignment could become a more exploratory, open-ended process of community engagement. In addition to meeting with their contacts at the organization about the immediate writing need, students could seek the perspectives of other stakeholders, especially the organization's clientele, on the larger community problem(s) that their project will address. Of course, this first requires an exploration of just who the various stakeholders might be and, specifically, what they have at stake. In the manner of Flower's intercultural inquiry, students could invite stakeholders to help them first define the problem (including its significance and underlying causes) and then deliberate about the best courses of action to address it. Instead of discussing the larger social and institutional conditions of the problem in a final reflection essay, students discuss them up front in the proposal. Instead of waiting to get input from targeted users later in the production process, students get this input at the beginning, when it can significantly shape their approach to writing, design, and other tasks.

In more recent courses, I've encouraged students to form focus groups with potential users and possibly other stakeholders as part of the research for their project proposal. A group of students proposing brochures to recruit people living with HIV/AIDS (PLWH/A) into an HIV services community planning council consulted literature about barriers to participation but also spoke with local service providers and, more important, several new PLWH/A members of the council. The students found that these two groups had markedly different perspectives from each other and from the students themselves. In another course, a team of students designing a website for a campus student organization not only formed a focus group of officers and members but also administered a questionnaire to the entire membership and even to potential members about what they would want and expect from such a website. Getting feedback about project objectives is especially important, as this can ensure that the project is responsive to a wider array of stakeholders. If students can't get this type of input from the very beginning, they can build into their proposal mechanisms for doing so as early as possible. Students can also discuss institutional or other barriers to getting stakeholders involved, barriers that their sponsoring organizations sometimes face as well.

Beyond revising the proposal assignment into an opportunity for early stakeholder engagement, this version would keep it flexible. This flexibility is also characteristic of the institutional critique developed by Porter et al., which requires that "changed practices be incorporated into the very design of the research project" (628). Students would use the proposal as a jumping off point for their ongoing usability efforts. Students would feel free to make substantial changes to their texts and processes as they become more aware of the complexities of their exigencies and better able to collaborate with other stakeholders, especially users. To illustrate, a team working with a Boys and Girls Club center thought their previous work with the agency and connection with its director would adequately position them to produce a persuasive leadership training video targeting teens. After they began developing their script, however, they realized how much they needed more direct input from the teens themselves, and they consequently appointed a group member to meet with the teens on a regular basis. In addition to initially talking to PLWH/A members of the HIV services planning council, the group producing recruitment materials later found it necessary to form of focus group of PLWH/A who were not already aligned with the council in any way and who therefore could approach the materials from a more novice standpoint.

One of the downsides of turning the proposal assignment into a more complex process is having less time in the semester to actually implement what has been proposed. Although students may have to take on more modest projects, these projects would hopefully be more responsive and empowering to users and other stakeholders. To supplement their contributions, students could propose that additional parts of the project or other, complementary projects be carried out by other groups or future classes (something I have them consider again toward the end of the semester). I'd rather students take on smaller but deeper projects than longer but shallower ones.

Another assignment that some technical communication courses require to aid students in their invention is a discourse analysis. The forum analysis developed by Porter is a type of this. Based on the concept of the discourse community, the discourse or forum analysis asks students to examine the background, social dynamics, values, and textual conventions of a particular discourse community as reflected in representative forums and texts (see Porter, *Audience*). Such an assignment can certainly aid students' rhetorical production and enculturation. It can also be limiting, however, when it is used as a tool for uncritical accommodation and narrowly focused on conventions within a discrete organization. One way to enhance this assignment is to expand its focus to the power dynamics among the various stakeholders in the project, including students themselves. Students could critique, in particular, how the organization's document design, production, and evaluation processes involve or fail to involve potential users and other stakeholders. They could assess whose values and perspectives get privileged in these processes and to what effects. In this way they could approach usability not just in terms of theories and approaches, but in terms of how ongoing power dynamics play out in actual, logistical practices. My student group producing job search materials for a halfway house for federal inmates began by mapping the power dynamics and constraints (mostly controlled by the Federal Bureau of Prisons) that regulated the inmates' job search process. This led the students to not only solicit the perspectives of inmates (through a liaison at the halfway house), but also to devise mechanisms that would give the inmates more agency in the job search process; the students contacted potential employers, for example, to find out who might hire inmates and to gather interviewing tips.

Expanding their discourse analysis outward, students could also connect such local exercises of power to broader social formations, such as patterns in agency-client relationships or the distribution of government funding to social service agencies. In the case of the job search project just mentioned, the students began to connect their observations about the constraints faced by inmates (e.g., short time frame, no phone or computer access) to a larger set of obstacles that make it difficult for inmates to reenter society successfully.

Unlike more pragmatic versions of the discourse analysis assignment, the main aim of this one would not be to help students successfully adapt to the status quo, but to help them strategize ways, however small, to ethically reconfigure the relationships in which their projects position them. As Sullivan and Porter define it, ethics is the art of "making careful decisions about how power relations are to be exercised in order to avoid domination of the other" (119). In *Opening Spaces*, they offer a heuristic—the advocacy chart—that could help students plan their ethical exercises of power. Such a chart could help student keep track of their positioning (vis-à-vis others involved), their roles, their advocacy plans, and evidence of their advocacy (183). It could also help them determine where and how they have the best chance to make an impact.

A third assignment that I have revised through an injection of cultural studies is the final report. Before I began experimenting with cultural studies, I ended my service-learning technical communication courses with a final evaluation report assignment. In this report students reflected on what they learned from the project, assessing both the process they underwent and the text(s) they produced. Their audience for the report consisted of the class and me. More recently, I have reconfigured this assignment as an "action plan," to borrow a term from Porter et al. This assignment, which could also be classified as a recommendation report, asks students to recommend ways to ethically implement, revise, or build on their projects. Students can address their recommendations to me, future groups of students (who will pick up where they left off), their sponsoring organization, and even other community stakeholders. Although students write the action plan at the end the semester, it is based on their semester-long assessment of their project and the problems to which it responds; in this way, the action plan is an extension of the project proposal. Like the proposal, the action plan enables students to channel their critical assessment into sustainable ethical action, and it asks them to move beyond description and evaluation to articulate new or revised goals and recommendations for obtaining them.

I usually ask students to make two sets of recommendations. The first set focuses on ways to implement, revise, or extend their projects in order to ensure their long-term sustainability. Students can start by suggesting ways to reproduce and distribute their text(s) (building on this discussion in their project proposal). To enable their video to circulate more widely, the video group proposed that the Boys and Girls Club make it accessible in various formats (e.g., VHS, DVD, the Web). Through their dialogue with project stakeholders, the group that produced HIV/AIDS community planning brochures identified new sites and special events at which the brochures could reach their PLWH/A audience. Such recommendations help students consider the life cycles of their texts in more concrete ways. Students who were unable to conduct user testing on their project (perhaps because they had limited access to their audience) can also recommend procedures for doing so after the semester is over and the project is in use. The student group producing a website for a campus organization left their sponsors with user testing guidelines and tools to implement after the site was fully operational. Even if students are able to generate feedback from user testing, they do not always have the time or expertise to incorporate all of their planned revisions at the end of the semester. The action plan gives them the opportunity to pass these revision ideas on to others (e.g., the organization's staff, future groups of students) who can incorporate them later.

The action plan's other set of recommendations focuses on ways to extend or complement students' projects. Students working on part of a large-scale project that will span multiple semesters can recommend directions for building on their work. A group that produced a set of staff procedures for a disabilities agency recommended that a future group collect samples of documents discussed in the procedures and add these as appendixes. This group also left future groups with their design template, style guide, and suggestions for coherently combining their section of the procedures with new sections.

The second set of recommendations can also focus on new, complementary projects that respond to the problem in alternative ways or that address another angle of the problem or its causes. This requires a reexamination of the problem, of course, something that students are likely better equipped to do after working on the problem for most of the semester. Asking students to consider alternative projects can help them put their project in perspective and recognize that it is only one part of a larger set of solutions. Sometimes students propose projects that closely complement their own. The group producing HIV/AIDS community planning recruitment brochures recommended that the agency and future student groups produce additional texts—such as PowerPoint slides, flyers, and a new page on the planning body's website—that could be used to complement their own.

Students may also identify complementary projects through their ongoing reflection. After studying a four-part special report on Living with Disabilities published by our regional newspaper, the *Orlando Sentinel*, several students working on a database user manual for a disabilities agency reflected on the need for educating others about the local demand for disability services and the funding limitations faced by their sponsoring agency. In their action plan, these students recommended that future groups of students produce materials to aid the agency's advocacy efforts. The next semester, another group of my students developed an advocacy manual that the agency can give to parents and other stakeholders to get them involved in advocacy campaigns. This was a good example of how the action plan can help sustain a long-term service-learning partnership.

Finally, students' action plans can recommend to their organization and future students ways to build on or learn from their collaborative *processes*, including their mechanisms for user involvement and their methods of collaborating with each other. Students can recommend (especially to other students) specific mechanisms for improving client involvement (through focus groups, earlier user testing, etc.) in text development, assessment, and revision processes. A group working on mental health brochures for the university's counseling center recommended that the center hold question-and-answer sessions with students and their families at the beginning of each semester to seek their input about needed services. The group producing a section of an organization's office procedures recommended to future students mechanisms for getting a wider array of staff members involved in the early development of new sections; through its action plan, this group helped other groups avoid some of the pitfalls that it faced. In this way, the action plan assignment helps students teach each other about the logistical and sociopolitical practices of usability and collaboration.

The action plan assignment draws on cultural studies impulses to enrich and sustain service-learning in several key ways. First, it helps students position their work in a larger context of responses to a communally defined and located problem. Second, in addition to calling for a more humble stance toward their projects, the action pushes students to channel their critical reflection into action, action that address the ethics of their work along the larger trajectory of its circulation and transformations. Perhaps most important, as students' action plans are taken up by the agency and the next course's students to negotiate new service-learning projects, these action plans help sustain a mutually empowering partnership and dialogue.

In all of these senses, the action plan, like my revised proposal assignment, is built on the cultural studies notion of articulation, which can be defined as the ongoing process by which temporarily coherent structures (in this case technical texts) are produced out of shifting, nonessential linkages of various conditions, including identities, organizational practices, community needs, economic resources, and cultural values. Students' work is taken up and rearticulated in response to shifting sets of conditions; at the same time, this work becomes part of the network of possible conditions that enable new articulations. I next want to experiment with using the action plan to build ongoing, sustainable projects across multiple classes with the same group of students. The service-learning based civic leadership programs at the University of Illinois at Chicago and Portland State University are good examples of curricular models that would enable such an extension of student-driven sustainability.

As the assignments I have been discussing show, cultural studies does not require us to abandon the pragmatic activities and assignments that form the backbone of many service-learning or project-based technical communication courses. It does, however, challenge us to revise and expand these activities and assignments to more critical and activist ends. This way cultural studies can help us fulfill two of the main goals of service-learning—to promote critical reflection about civic responsibility and to enact this responsibility in projects that are responsive and empowering to others in one's community.[2] When inflected by cultural studies, service-learning approaches to technical communication can better live up to their transformative potential.

NOTES

1. Lately I have begun to rethink my basic model of service-learning as well as my assignments. Instead of having students work on projects designated by specific community and campus organizations, I want to try a more open-ended model in which students begin with an issue rather than an "assignment" or specific client. Instead of adjusting to project parameters largely predetermined by a sponsor, students would first explore an issue with other stakeholders in a kind of needs assessment. This would not be completely open-ended, as students would still need to fit their assessment within a service-learning framework; it would need to meet a community need and reinforce course content and objectives, for starters. A needs assessment can be a particularly rich assignment in that it invites a sustained dialogue with others and calls for a complex set of responses. Service-learning projects need not be framed as "solutions" that are developed, produced, and implemented in one fell swoop; indeed, this kind of approach often leads to unsustainable interventions. If one class performs a preliminary assessment or analysis of an issue, subsequent classes can continue the work of developing specific, manageable follow-up projects.

2. Many corporations have adopted the language of civic responsibility, stakeholder engagement, and community partnership in their public relations responses to pressure from shareholders, NGOs, and other groups. This rhetorical appropriation, often unaccompanied by robust efforts to engage stakeholders as partners, makes me worried that even the approach I am advocating here could be subsumed into a pragmatist project of preparing students to be effective corporate workers who can talk the talk of civic engagement.

WORKS CITED

Anson, Chris M. "On Reflection: The Role of Logs and Journals in Service-Learning Courses." *Writing the Community: Concepts and Models for Service-Learning in*

Composition. Ed. Linda Adler-Kassner, Robert Crooks, and Ann Watters. Washington, D.C.: American Association for Higher Education and NCTE, 1997. 167–180.

Bickford, Donna M., and Nedra Reynolds. "Activism and Service-Learning: Reframing Volunteerism as Acts of Dissent." *Pedagogy* 2.2 (Spring 2002): 229–252.

Blyler, Nancy. "Taking a Political Turn: The Critical Perspective and Research in Professional Communication." *Technical Communication Quarterly* 7.1 (1998): 33–52.

Bowdon, Melody, and J. Blake Scott. *Service-Learning in Technical and Professional Communication*. New York: Longman, 2003.

Carey, James W. "Overcoming Resistance to Cultural Studies." *What Is Cultural Studies?* Ed. John Storey. London: Arnold, 1996. 61–74.

Crowley, Sharon. *Composition in the University: Historical and Polemical Essays*. Pittsburgh: U of Pittsburgh P, 1998.

Deans, Thomas. *Writing Partnerships: Service-Learning in Composition*. Urbana, IL: NCTE, 2000.

Dubinsky, James M. "Service-Learning as a Path to Virtue: The Ideal Orator in Professional Communication." *Michigan Journal of Community Service Learning* 8 (Spring 2002): 61–74.

Flower, Linda. "Partners in Inquiry: A Logic for Community Outreach." *Writing in the Community: Concepts and Models for Service-Learning in Composition*. Ed. Linda Adler-Kassner, Robert Crooks, and Ann Watters. Urbana, IL: AAHE and NCTE, 1997. 95–117.

Grossberg, Lawrence. *Bringing It All Back Home: Essays on Cultural Studies*. Durham, NC: Duke UP, 1997.

Hall, Stuart, and Lawrence Grossberg. "On Postmodernism and Articulation: An Interview with Stuart Hall." *Journal of Communication Inquiry* 10.2 (1986): 45–60.

Hea, Amy C. Kimme. "Developing Stakeholder Relationships: What's at Stake?" *Reflections: Writing, Service-Learning, and Community Literacy* 4.2 (Winter 2005): 54–76.

Henry, Jim. "Writing Workplace Cultures." *College Composition and Communication Online* 53.2 (2001): www.ncte.org/ccc/2/53.2/henry/article.html.

Henson, Leigh, and Kristene Sutliff. "A Service Learning Approach to Business and Technical Writing Instruction." *Journal of Technical Writing and Communication* 28.2 (1998): 189–205.

Herndl, Carl G. "Teaching Discourse and Reproducing Culture: A Critique of Research and Pedagogy in Professional and Non-Academic Writing." *College Composition and Communication* 44 (1993): 349–363.

Herzberg, Bruce. "Community Service and Critical Teaching." *College Composition and Communication* 45.3 (1994): 307–319.

Huckin, Thomas N. "Technical Writing and Community Service." *Journal of Business and Technical Communication* 11.1 (January 1997): 49–59.

Johnson, Richard. "What is Cultural Studies Anyway?" *Social Text* 6 (1987): 38–80.

Katz, Steven B. "The Ethic of Expediency: Classical Rhetoric, Technology, and the Holocaust." *College English* 54.3 (March 1992): 255–275.

Kynell, Teresa C. *Writing in a Milieu of Utility: The Move to Technical Communication in American Engineering Programs, 1850–1950.* Norwood, NJ: Ablex, 1996.

Liu, Goodwin. "Origins, Evolution, and Progress: Reflections on a Movement." *Metropolitan Universities: An International Forum* 7.1 (Summer 1996): 25–38.

Longo, Bernadette. *Spurious Coin: A History of Science, Management, and Technical Writing.* Albany: SUNY P, 2000.

Matthews, Catherine, and Beverly B. Zimmerman. "Integrating Service Learning and Technical Communication: Benefits and Challenges." *Technical Communication Quarterly* 8.4 (Fall 1999): 383–404.

Porter, James E. *Audience and Rhetoric: An Archaeological Composition of the Discourse Community.* Englewood Cliffs, NJ: Prentice Hall, 1992.

Porter, James E., Patricia Sullivan, Stuart Blythe, Jeffrey T. Grabill, and Libby Miles. "Institutional Critique: A Rhetorical Methodology for Change." *College Composition and Communication* 51.4 (June 2000): 610–642.

Rehling, Louise. "Doing Good While Doing Well: Service-Learning Internships." *Business Communication Quarterly* 63.1 (March 2000): 77–89.

Salvo, Michael J. "Ethics of Engagement: User-Centered Design and Rhetorical Methodology." *Technical Communication Quarterly* 10.3 (2001): 273–290.

Schultz, Aaron, and Anne Ruggles Gere. "Service Learning and English Studies: Rethinking 'Public' Service." *College English* 60 (1998): 129–249.

Scott, J. Blake. "Tracking Rapid HIV Testing Through the Cultural Circuit: Implications for Technical Communication." *Journal of Business and Technical Communication* 18.2 (2004): 198–219.

Sullivan, Dale L. "Political-Ethical Implications of Defining Technical Communication as a Practice." *Journal of Advanced Composition* 10.2 (Fall 1990): 375–386.

Sullivan, Patricia, and James E. Porter. *Opening Spaces: Writing Technologies and Critical Research Practices.* Greenwich, CT: Ablex, 1997.

Chapter Eleven

Designing Students

Teaching Technical Writing with Cultural Studies Approaches

Katherine V. Wills

> The first rule of method is that we shall have no rule about what constitutes knowledge.
>
> —Bruno Latour, *Science in Action*

To better prepare technical writing students for the postmodern workplace that rewards innovation and creativity, technical writing curricula and classrooms that feature elements of cultural studies approaches can increase students' awareness of cultural/hegemonic systems and rule-governed contexts, including the college classroom. Teachers can assist students' transit from reproducing culture unconsciously to producing culture—technical documentation—consciously, knowledgeably, and effectively by also investigating and analyzing sociopolitical power. In "An Approach for Applying Cultural Study to Technical Writing Research," Bernadette Longo confirms technical writing's historical affinity for scientism. By combining cultural studies with technical writing pedagogy, we facilitate classroom conversations on authority, culture, and power. In order for graduate-level technical writing students to move from reproducing culture unconsciously to producing culture consciously, they are introduced to research methods and writing strategies based on a sociopolitical analysis of power. Conversely, cultural studies can benefit by its association with technical writing. Innovation often entails the amalgamation of ideas, materials, contexts that were previously thought to be unthinkable, incompatible, or impossible. The joining of cultural studies and technical communication fuses social theory with technical praxis into hybrid possibilities. Cultural studies theorists can test social hypotheses in technical communication classrooms. Cultural studies can enter more diverse disciplinary discussions through technical writing.

I frame my argument with Pierre Bourdieu's *Homo Academicus*, in which he connects the accumulation of academic capital to academic resistance and to methodological change. Bourdieu assumes a spectrum that situates "cultural production" mostly with those professors who conduct research. At the other end of the spectrum, professors focused on teaching do the work of cultural reproduction within his French educational system (109). Bourdieu explains the persistent accumulation of academic capital that reifies underlying extant hierarchies. Academic capital manifests itself as power through the establishment of hierarchies or "practical taxonomies" (194). Bourdieu describes academics as the species "*Homo Academicus*, supreme classifier among classifiers" (i). Regarding taxonomies, his study reveals that Althusser, Barthes, Foucault, Derrida, Deleuze, and others subsisted in marginal positions in the French university system that prevented them from doing research. Their status forced them to cultivate strong relationships outside the university, thus shifting their perspectives and allegiances (*Preface*). According to Bourdieu, these shared assumptions of order and meaning by academics pass along knowledge until more refined or historically relevant taxonomies and definitions take their place. With the advent of postmodern sensibilities, those who integrate cultural studies with technical writing instruction and methods must confront entrenched Victorian notions of knowledge making, thus forcing those comfortable with teaching and research methods (and their instruction in graduate schools and technical programs) to rethink the goals and structure of outdated technical writing curricula based on hierarchical taxonomies from nineteenth century paradigms.

Cultural studies approaches in technical communication permit teachers and students to access the ambient cultural and social paradigms of their age, not the previous age. At this time, highly interconnected, global, and multicultural paradigms are most appropriate for capturing the most effective communication strategies in the classroom. Teachers and innovators would do better to be receptive to the conceptual tools latent in their time—these times. Technical communication, too, is a form of taxonomic organization of materials, thoughts, and processes. The most dramatic scientific taxonomic academic exercise of our late twentieth- and early twenty-first century era is the Human Genome project in gene categorization. This project rivals categorization such as Linnaeus's ordering of botany, Darwin's ordering of species, and Freud's attempt to order the human psyche. Configuring the human genome relies on web and matrix patterns of organization made more accessible to the imagination by digital technologies. Darwin relied on linear progression to "see" natural selection and human evolution; thinkers now rely on three-dimensional (or more) matrices to capture the essence of the human gene pool.

Bourdieu describes why new forms of classification are needed in the postscript to *Homo Academicus*, "Categories of Professorial Judgment." He concludes: "Thus we shall find at work *academic forms of classification* which, like the 'primitive forms of classification' mentioned by Durkheim and Mauss, are transmitted in and through practice, beyond any specifically pedagogical intention" (197). The forms of classification were and continue to be reproduced by the academic system without regard for the changing needs of society. They are intellectual products of habit, not of contemporary human life: they are "an enterprise of cultural production for the purposes of reproduction" (224). For example, in his postscript to *Homo Academicus*, Bourdieu observes that the documentation of 154 files from a *premiere superieure* school in Paris seemed to academically classify empirical information about the students. On closer examination, Bourdieu states:

> This means that the official, specifically academic, taxonomy, which is objectified in the form of a series of adjectives, fulfills a dual and contradictory function: it allows the operation of a social classification while simultaneously masking it; it serves at once relays and screens between the classification at entry, which is overtly social, and the classification at exit, which claims to be exclusively academic. (201)

Bourdieu insists that academic power comprises the ability to "influence on the one hand expectations—themselves based partly on a disposition to play the game and on investment in the game, and partly on the objective indeterminacy of the game—and on the other hand objective probabilities—notably by limiting the world of possible competitors" (*Homo* 89). In other words, academic authority is premised on inveterate commitment to hierarchical authority, especially in Bourdieu's time. Academics are in their positions to reproduce themselves and the academy. Following Bourdieu's lead and then inverting it for today's postmodern workplace, expectations for technical writing should compete with instantiated assumptions about the nature of methods used in teacher training for technical writing in the academy, which are often situated in English departments, or business and technical programs.

As teachers, we need to bring to our pedagogy an awareness of the anti-intellectualism operating within universities as we try to teach technical writing with an eye toward the social patterns inherent in each technical communication event.

> The secret resistance to innovation and to intellectual creativity, the aversion to ideas and to a free and critical spirit, which so often

> orient academic judgments, as much as the viva of a doctoral thesis or in critical book reviews as in well-balanced lectures setting off neatly against each other the latest avant-gardes, are no doubt the effect of the recognition granted to an institutionalized thought only on those who implicitly accept the limits assigned by the institution. (95)

Rather than viewing a cultural studies application to the methods of teaching technical writing, this intersection of cultural studies and technical writing instruction can be seen as situated in Bourdieu's familiar dichotomy or binary of "two different modes of production and reproduction of knowledge and, more generally... two ways of envisaging the successful man [*sic*]" (*Homo* 58). These two views may be between scientific and social competence, the researcher and the clinician, and the canonical and the marginal. Teaching methods with an awareness of the power and social context present in every writing or communication act—no matter how objective, reliable, or valid the act may seem from a hyperpragmatist or positivist stance—will heighten students' rhetorical sophistication. From these foundations, we can further contribute to student preparation by expanding teaching that is contextualized in cultural studies and the twenty-first century workplace paradigms. Technical communication pedagogy in general, and graduate training programs that directly or indirectly prepare future teachers of technical writing, need to access and foreground the social, cultural, and environmental contexts of teaching technical writing.

Students (and teachers) should be disabused of the notion that merely replicating formats, tables, data, and terms for a grade of C or better will provide job security. Many students with work experience have internalized the belief that much of college coursework including technical writing are "hoops" that students jump through to get to the "real" world and pay. Anecdotally, "hoops" might include assignments that seem irrelevant to students, a portfolio method that requires revisions and process, a-contextual exercises, and depending on the student, anything from rote memorization to calls for critical thinking and Socratic methods. Students expect to transform noneconomic capital such as education, social class, lineage, and even health into academic cultural capital that provides economic capital, access, and leverage toward increased power, money, and authority in the social world.

Teaching methods for technical writing have been mostly pseudotransactional. Courses have relied on textbooks and assignments whose primary audience is the college writing instructor, not technical readers or writers from the workplace. Teachers of technical writing courses at the undergraduate level are typically adjunct lecturers or graduate students who have had little authentic experience with workplace writing. This may explain, in part, the reliance on unexamined adherence to authoritative texts, slavish reliance on extant taxonomies, and a dependency on fossilized formulaic methods of teaching and research. Documents created by students from the assignments are decontextualized imitations of what might occur in the workplace. Such assignments focus on supplying students with a standardized educational experience. Students recognize that their assignments are somewhat contrived.

Teacher training becomes less pseudotransactional with the advent of internships, service-learning, and real-world writing assignments that situate the educational experience outside the academic culture and within various workplace cultures. Including these additive sociocultural workplace perspectives in the teaching of technical writing provides ways of interpreting and designing technical writing assignments outside standard academic hierarchies that have dominated technical writing.

Pedagogical Strategies for Combining Cultural Studies with Technical Communication

How can teachers introduce cultural studies to their technical writing and research methods pedagogy? Within graduate writing programs, academics continue to pass along the shared assumptions that have been situated within established hierarchies. One assumption is that the teaching of technical writing need not be concerned with the questioning of power or authority within systems. A cultural studies perspective to technical writing could benefit from just such an adjustment to pedagogy. According to Longo, cultural studies approaches to technical writing are antidisciplinary, suggesting a plethora of possible pedagogical strategies. While cultural studies cannot claim the alleged same objectivity of scientism, it can, instead rely on the "analysis of situatedness of technical writing practices within a culture." Longo offers these six (of many) channels for knowledge making, which can be applied to technical communication pedagogy: analyzing the historical context; analyzing cultural context; critiquing gaps between dominant culture and field practices; recognizing the effects of the researcher/teacher on the knowledge-making process; and pinpointing discursive artifacts of language use (66). Cultural studies approaches can be combined with technical communication pedagogy in the following ways. A cultural studies technical writing classroom functions as an antidisciplinary and multi-theoretical umbrella to incorporate a variety of learning strategies and concepts.

Dragga, Kells, Dobrin and Weisser, Scott, Wilson, and Longo provide five of many possible methods that accentuate how applying cultural studies pedagogy to technical communication can be applied in the technical communication classroom while preparing students for socioculturally imbricated document production. Students who are introduced to a variety of strategies based in cultural studies can better recognize ideological agendas and power relationships among writing audiences. Teachers can use examples such as the ones below to argue to students that being receptive to the use of diverse cultural studies approaches can enhance technical writing and workplace effectiveness. Having cultural, social, and contextual knowledge of power in a given writing context can produce better documents.

Documents are not merely scientific or comprised of quantifiable information for representation in charts, graphs, or figures. Cultural interpretation and knowledge drives the most effective technical product. With this recognition, students should bring a heightened awareness of writing purposes and audiences to their products. Even the course syllabi could be treated as a document of the academic workplace and a "real world" document.

Syllabi

Syllabi serve as a handy model for having students examine power in the workplace. If the instructor confronts students the first day of class with the syllabus as a workplace document, they might quickly confront their own complicity, or lack thereof, in academic power hierarchies. By using the syllabus as a first example of document production that reveals, relinquishes, or negotiates power in the classroom, instructors foreground authority (academic, workplace, or institutional). Using the syllabus is particularly effective because of its effect on student workloads and grades. The oft-cited division between "real-world" and academic instruction is blurred by using the syllabus as workplace document. Furthermore the classroom transforms into a "real-world" workplace where document production has consequences and immediate relevance to students.

Curricula

In another example, Sam Dragga shows how applying the cultural paradigm of Confucian ethical understanding is pivotal to successful marketing of Kellogg Company's Coco Puffs™ cereal in the Chinese marketplace. Implicit in Dragga's study is the idea that teachers and researchers of technical writing (especially intercultural communication) should apply a variety of appropriate cultural lenses, in this case a lens of Confucian ethics, for more effective technical communication. He suggests that offering culturally based frames fortifies technical communication by more accurately analyzing "the moralities driving the communication processes of specific civilizations" (379).

Dragga analyzes how technical writers and designers infuse the rhetorical messages on a Kellogg cereal box with an understanding of Chinese and Confucian values so as to appeal to the Chinese consumers. The package was designed with cultural understanding of the primacy of tradition, family, and food among Chinese consumers. Specifically, Dragga notes the use of the color red "because red in China is the color of good luck." Additionally, Kellogg emphasizes that it is a sincere and family-centered company by displaying its "all families" symbol. Kellogg designed the outside of the cereal box with a Coco Puffs™ monkey (a culturally revered animal) cartooning as a teacher extolling the virtues of nutrition. Applying the technical communication textbook templates alone, Kellogg's designers would not have investigated these value-situated, culturally resonant inflections of cereal box design.

In her session titled "Teaching Intercultural Communication Principles in the Technical Writing Classroom" at the 2003 Modern Language Association Conference, Michele Hall Kells describes an exercise in a technical writing class that linked document production processes with heightened connections to "students' physical and cultural environments." Kells asked students to employ reflective practices in journaling and collaborative work. Students drafted grant proposals for international projects, some of which were funded. Students also created websites when targeting an international audience. In her pedagogy, Kells foregrounds cultural contexts of pedagogical and rhetorical choices in all stages of the writing process. She stated: "Technical writing students need to learn effective ways to attend to the 'noise' (not dismiss culture as something to be 'ignored or expunged')." Graduate students could be asked to teach technical writing using texts drawn from technical communication discourse such as the loss of the Columbia, nuclear reactors, jury instructions, tax accounting (Bazerman and Paradis) and events such as September 11, 2001.

Creative teaching takes time and risk. If a pedagogical goal is to assist students in becoming better readers and writers of power issues surrounding document production in the workplace, the reliance on textbooks and formulas will have to take a temporary backseat to the instructor's vision and students' effectiveness. By engaging students in such socially conceptualized technical writing discourses, we socialize technical writers and communicators into a community.

Venues and Service–Learning

Assignments that integrate off-campus writing internships or collaborative work programs for graduate students are fertile ground for contrastive critiques of institutional power. Not only can students practice technical writing within transactional workplace, they can actualize the teaching adage that learners learn best by doing:

If you tell me, I forget.
If you show me, I remember.
If you involve me, I understand.

Involving technical communications students in community projects with stakeholders not only enhances understanding, but also elicits probing reflection of power structures that determine technical writing projects, especially when used along with critical inquiry questions.

In "Breaking Ground in Ecocomposition," Sid Dobrin and Christian Weisser offer ecological literacy approaches for technical communication pedagogy that challenge students and teachers to recognize interconnected and synergistic systems of social activity that influence any communication event. Marilyn Cooper's ecological model posits that pedagogies that foreground the sociocultural nature of writing "signal a growing awareness that language and texts are not simply the means by which individuals discover and communicate information, but are essentially social activities, dependent upon social structures not only in their interpretive but also in their constructive phases" (Cooper as quoted in Dobrin and Weisser 568). And in this collection, Blake Scott argues that service-learning has at least as much in common with cultural studies as pragmatism, phronesis, and civic ethos. Specific assignments are modified to capture ethical concerns. For example, Scott cites Hensen and Sutcliff as they inquire of students, "What social conditions have created the need for the nonprofit you serve?" Scott's technical communication assignments have included proposals that serve as flexible heuristics rather than fixed boiler-plated documents, discourse analyses that examine power dynamics within organizations, and actions plans that utilize the cultural studies notion of "articulation."

Alternative Theoretical Framing

Greg Wilson suggests a postmodernist frame that draws on the writings of David Harvey and Robert Reich (see also Johnson-Eilola). He deflects arguments that applying a "structureless approach to the structured description of structured systems" may seem counterintuitive (72). Instead, he argues that this wrenching of assumptions about technical communication could loosen the following: modernist thinking patterns that ignore postmodern realities; rule-governed definitions; rigid classifications; rigid notions of facts and truth; and reliance on linear hierarchal representations (74) with technical communicators as mere "scribes and translators" (96).

Wilson's assumptions about what constitutes knowledge in the 2001 technical communication classroom differs from what colleagues Couture et al. envisioned in 1985. In "Building a Professional Writing Program Through a University-Industry Collaborative," Couture et al. assumed that two major purposes of a writing curriculum are to improve student "skills" while "unifying diverse perspectives" of writing instructors and researchers (Couture et al. 409). The authors suggest a "step by step process" of instruction focusing on research and teaching methods of the reproduction of measurable outcomes for a like-minded audience (423). Skills-based pedagogies aimed toward unified perspectives have their place in technical communication pedagogy. Increasingly, however, diversity, innovation, and space/time flexibility permeate workplace production; these concepts need to be incorporated and reflected in technical documents.

Wilson concludes that technical communication pedagogy permitted little agency to the student technical writer; however, in the postmodern workplace, students who are educated to be self-actualizing symbolic analysts in both modernist and postmodernist veins will have technical communication preparation superior to "scribes and translators."

Modernist workplace realties are not as closely in tune with post-Fordist jobs and postmodern "flexible accumulation" that move away from production to service (79). "In terms of pedagogy," Wilson writes, "courses that focus on technical communication as symbolic analysis may better prepare students to be corporate citizens. . . . The (post)modern work world is interested in innovations and in synthesis of overwhelming amounts of information" (86). Incorporating postmodern perspectives in classroom content does not necessarily entail adding or removing conventional textbooks and related skills. Technical communication students simply would need to have more exposure to the following applications and processes:

Abstraction: The notion that "Workers never make decisions with perfect information" prepares students to think creatively and with an eye toward their social complicity when producing documents (87). Exercises in abstraction assist students in conceptualizing alternative answers and avoiding assignments with instructor-stipulated parameters and well-worn answers. This allows students to figure out what they have to do when they have no directions and no textbook.

Systems thinking: Systems thinking replaces compartmentalized thinking with weblike or interrelated thinking, encouraging "more complex viewpoints in their [student] writing" (89). Wilson uses Johndan Johnson-Eilola's example of a class assignment using maps (see also Longo 61). The maps show "differences, contradictions, and cultural assumptions" (89). As students examine the maps, they realize that their conventional notions of maps as manifestations of objective reality are flawed.

Experimentation: "Throwing out the rules" and not following modernist experimentation as a "structured, linear, rule-bound procedure" liberates the better aspects of postmodern pedagogy (Wilson 95). Wilson suggests assignments that highlight experimentation and focus on "the exploration of the interactions between participants, discourses, and contexts" (95).

Collaboration: Wilson's symbolic analysis in the postmodern workplace (academic or industrial) shows pandemic reliance on group and team efforts. Collaboration is already a standard strategy in some instructors' technical communication pedagogy. Research studies support the idea that collaboration facilitates students' acculturation into professional and workplace discourse communities (Thralls and Blyler 13).

Theorization: In *Spurious Coin: A History of Science, Management and Technical Writing*, Longo describes how Lester Faigley applied Foucault's archeology method in writing classrooms to bring to light "individual subject positions in the composition classroom because Foucault's work, and critical theory in general, provide a theoretical basis for recognizing institutional relationships and a vocabulary for discussing power/knowledge systems" (18). By applying Foucauldian archeology critical theory to the technical communication classroom, teachers explore how knowledge is legitimated or marginalized within institutions, relationships and classrooms (Longo, "Approach" 119; 61 in original).

DEALING WITH RESISTANT AGENTS WHEN COMBINING CULTURAL STUDIES AND TECHNICAL COMMUNICATION

How can instructors motivate resistant students who are more comfortable with hyperpragmatist teaching? This resistance to examining how power issues affect document production is not a personal affront to the instructor, but a cultural educational artifact. Dealing with students who resist changes in their comfortable ways of thinking is not unique to the teaching of technical communication. Nonetheless, students coming to a technical communication class from diverse or international systems of education that emphasize rule-governed behavior might also be more resistant as an audience than students in other disciplines, generally. Anecdotally, there are numerous stories of students in the sciences, such as engineering or medicine, who resist asking questions of their instructors or challenging familiar solutions when taking a required technical writing course (see also Dragga 377).

We see that resistance is also endemic to the professorate. Bourdieu argues that science and arts professors who bootstrapped themselves up from the lower or middle classes view their perceived success of rising to the upper classes as contingent on their college education and their adherence to rule-governed behavior and production (52 *Homo*). Bourdieu argues that professors, too, tend to "reinvest totally in the institutions which have so well rewarded their previous investments, and they are very little inclined to seek power other than university power" (52). Professors who come from bourgeois origins (Bourdieu mentions lawyers) combine their academic power with political or business power. Professors' "intellectual dispositions" influence the pedagogical parameters—the risks—that the professors are willing to take in the classroom. Many professors and students of technical writing owe their academic success first to the power structure of the academy and resist alternative venues of power.

Professors and students break free of the unexamined reproduction of structures of dominance not by replacing them with other structures; rather, agents (students and teachers) recognize their complicity with extant power structure and choose to differentiate themselves—or not. Without the recognition, however, agents may simply and unconsciously substitute one set of rules for another.

In teaching and preparing graduate students to teach technical communication with cultural studies approaches, what constitutes knowledge should be presented in as many possible perspectives as feasible in a course. Recognizing how power operates in the ongoing production and reading of documents will be in any given moment contingent on the preparation and imagination of all students and teachers involved. As researchers, teachers, and theoreticians apply cultural studies in inventive ways, cultural studies will grow along with the practices and perspectives of the new millennium.

Works Cited

Bazerman, Charles, and James Paradis. *Textual Dynamics of the Professions*. Madison: Wisconsin UP, 1991.

Bourdieu, Pierre. *Homo Academicus*. Stanford: Stanford UP, 1984.

Couture, Barbara, Jane Rymer Goldstein, Elizabeth L. Malone, Barbara Nelson, and Sharon Quiroz. "Building a Professional Writing Program Through a University–Industry Collaborative." *Writing in Non-Academic Settings*. Ed. Lee Odell and Dixie Goswami. New York: Guilford, 1985. 408–423.

Dobrin, Sidney I., and Christian Weisser. *Natural Discourse: Toward Ecocomposition.* Albany: SUNY P, 2002.

Dragga, Sam. "Ethical Intercultural Technical Communication: Looking Through the Lens of Confucian Ethics." *Technical Communication Quarterly* 8.4 (1999): 365–381.

Johnson-Eilola, Johndan. "Relocating the Value of Work: Technical Communication in a Post-Industrial Age." *Technical Communication Quarterly* 5.3 (1996): 245–270.

Kells, Michele Hall. Modern Language Association Conference. Association of Business Communication Session. San Diego, 29 Dec. 2003.

Latour, Bruno. *Science in Action: How to Follow Scientists and Engineers Through Society.* Cambridge, MA: Harvard UP, 1987.

Longo, Bernadette. "An Approach for Applying Cultural Study to Technical Writing Research." *Technical Communication Quarterly* 7.1 (1998): 53–74.

———. *Spurious Coin: A History of Science, Management, and Technical Writing.* Albany: SUNY P, 2000.

Odell, Lee, and Dixie Goswami. *Writing in Non-Academic Settings.* New York: Guilford, 1985.

Scott, J. Blake. *Risky Rhetoric: AIDS and the Cultural Practices of HIV Testing.* Carbondale, IL: Southern Illinois UP, 2003.

Sullivan, Patricia, and James E. Porter. *Opening Spaces: Writing Technologies and Critical Research Practices.* London: Ablex, 1997.

Thralls, Charlotte, and Nancy Roundy Blyler. "The Social Perspective and Professional Communication." *Professional Communication: The Social Perspective.* Ed. Nancy Roundy Blyler and Charlotte Thralls. London: SAGE, 1992. 1–34.

Wilson, Greg. "Technical Communication and Late Capitalism: Considering a Postmodern Technical Communication Pedagogy." *Journal of Business and Technical Communication* 15.1 (2001): 72–99.

Afterword

Diana George

> Trying to protect his students' innocence
> he told them the Ice Age was really just
> the Chilly Age, a period of a million years
> when everyone had to wear sweaters.
>
> —Billy Collins, "The History Teacher"

Several years ago when I was away for a one-year stint as a visiting professor of professional communication at another university, I began my seminar by asking students to interview people in the workplace, asking them to talk about their communication practices. Through those interviews, students discovered that computer programmers as well as nurses and history professors spend much of their workday writing. One student interviewed a bar owner who logged at least fifteen different kinds of writing ranging from rules for employees and customers to newspaper ads to brochures to letters to the town council on behalf of the local business owners' association. Another student even used the assignment to try to trick his father into telling him what he really does for a living. (He never found out except that it had something to do with the government, it did involve a lot of writing, and his father will not talk about it.)

Mine is a fairly typical assignment in professional communication. Many of my readers will recognize it and have very likely used it in their own classes. After all, researchers have been tracking workplace communication practices for quite a long time now, so my students' findings were interesting but not surprising. The interviews were simply a way for us to get ourselves into the business of a course in professional communication. The class had been reading *Writing and Reading Mental Health Records: Issues and Analysis*

(Reynolds et al.), and so I invited a colleague from Counseling Services to talk to them about communication practices in mental health.

I cannot say that I expected more from my own informant than my students had gotten from theirs—that the uses of writing in the workplace are multiple and varied. Perhaps because mental health professionals are often bound by the language of the *Diagnostic and Statistical Manual of Mental Disorders* (*DSM*), I expected the speaker to address the connection between the *DSM* and issues related to insurance, the law, and to medical referral and treatment. Instead, our speaker began by describing a rape case in which his clinical records were enlarged to the size of the wall and used to free a defendant the speaker was quite sure was guilty. Our visitor spoke with regret and, he said, some measure of embarrassment over the uses that had been made of his notes, which he had offered only to help prove the plaintiff's case. Instead, the professional record, brought into court by the accuser, had been enlarged, projected, and read over and over again daily in the court. Nowhere in these notes had he ever written that the client had been raped, that she had said she had been raped, or even that she had said she feared being raped.

"I had no doubt that her story was true," the speaker told my class, "but my notes were careful. I was trying not to put too much into the record. I thought I was protecting my client and myself." He dropped his head as he described how it felt to see his words up there and to know that, in the end, they said nothing at all. They were professional. They followed all good standards for mental health records, but they never showed a frightened woman or took note of the man she feared. In the end, they were, he said, truly useless as records of anything that actually happened.

If I read his fine contribution to this collection correctly, Jeffrey T. Grabill might well ask at this point in my story, in what way does cultural studies offer a significant intervention into a situation like this one? How might a critical reading of this text offer a "meaningful alternative" to actual practice in the writing and uses of mental health records? It is an important question—one raised throughout this volume and one constantly raised before cultural studies, which has always aspired to be more than just another academic discipline.

To address that question we might begin with the simple fact that language, as Jim Berlin often wrote, is never innocent. Language matters. The uses of language matter. The men and women who make rules about language and language use matter—especially in the courts, in the city halls, in our schools and colleges, in hospitals, and on the streets.

Cultural studies does offer a theory of critical analysis for language like this. More than textual analysis, however, cultural studies asks us to inhabit the spaces in which this language is created, made possible, and relied on.

Drawing from roots in cultural anthropology, cultural studies would demand that we not simply read the *DSM* and note its inadequacy as a language, but understand it as a system that works against both the mental health worker and the patient or client at the same time that it provides a protective frame within which courts and insurance companies can comfortably perform their work.

Cultural studies would certainly begin by examining in what ways the language and form of mental health records, for example, both constrain and free the professional. In what way is the court scene an opportunistic space for that language, that profession? In what way does the mental health record shift from being an aid to the mental health professional to a piece of evidence offered against the patient? In what way does the mental health profession work in concert with the legal system to keep in check or contain mental health practices? Furthermore, in a case like this one, cultural studies would examine the space this professional found himself in: convinced that the defendant is guilty; sure the defendant has raped not one but several women; equally sure that his client was served badly by both the legal system that used his words to "prove" that she invented the rape and by a profession bound by the constraints of language (*DSM*) and fear (malpractice suits). He sees his words blown up to magnificent proportions, his notes as tall as a man's hand, placed in front of him (the witness) who can do no more than assure the court that these are, indeed, his words.

Yet, to go back to Jeffrey Grabill's question about how cultural studies actually intervenes in a way that makes possible different practices/different frames, I must—given my own hopes and limitations—plead pedagogy.

I opened this afterword with a passage from poet laureate Billy Collins's wonderful little poem "The History Teacher." The poem tells the story of a teacher too soft-hearted, too kind to tell the truth. In this teacher's world, "the War of the Roses took place in a garden," and the history teacher wonders seriously if he can make his students believe "that soldiers in the Boer War told long, rambling stories / designed to make the enemy nod off." I love this poem, partially because I understand that teacher and partially because it is so funny/so sad, but I also love this poem because it tells us so much about the possibilities and limits of the classroom.

Technical and professional communicators often learn their lessons in our classrooms. I know, of course, that classrooms cannot lay claim to single-handed social change. In fact, in the hands of a classroom dominated by standardized tests, one that teaches uncritically the proper forms, the accepted language, the prevalent genres, there is little likelihood that our instruction is about much other than the classroom work itself. A teacher who tries too hard to protect his or her students from the harshness of racism, sexism, economic, social, and political inequities, labor practices, legal bungling, and more, might create a safe space but I would argue it is not a terribly useful space for enacting change, for actually intervening.

In order to even think of intervention, practitioners need to notice that something is wrong, and so I place a great deal of hope in a classroom informed by the inquiry and reflection made possible through cultural studies approaches to our teaching. And that is where, I believe, this volume leaves us—from theory to research to pedagogy. It is a fine and natural place to end because, of course, it is the place where we begin.

Contributors

Elizabeth C. Britt is Associate Professor of English at Northeastern University. She is the author of *Conceiving Normalcy: Rhetoric, Law, and the Double Binds of Infertility* (Alabama, 2001), as well as several articles and book chapters.

Bradley Dilger is an assistant professor at Western Illinois University, where he teaches composition studies, new media, and professional writing. His research centers on the influence that "ease" bears on culture and technology, from nineteenth-century composition classrooms, to technical communication, to networked computing.

Diana George is Professor of Humanities at Michigan Technological University, where she teaches courses in composition pedagogy and theory, cultural studies and composition, visual representation, popular culture, and British literature. She is the author of *Kitchen Cooks, Plate Twirlers, and Troubadours: Writing Program Administrators Tell their Stories* (Heinemann, 1999), coauthor, with John Trimbur, of *Reading Culture: Contexts for Critical Reading and Writing* (5th ed., Longman, 2004), and coauthor, with Lester Faigley, Cynthia Selfe, and Anna Palacik, of *Picturing Texts* (Norton, 2004).

Jeffrey T. Grabill is Associate Professor of Rhetoric and Professional Writing, Director of Professional Writing, and Co-Director of the Writing in Digital Environments Research Center at Michigan State University. The author of *Community Literacy Programs and the Politics of Change* (SUNY, 2001), Grabill is currently examining the use of advanced information technologies and professional communication for civic purposes. Several of his articles have won national research awards.

Jim Henry is Associate Professor of English at the University of Hawai'i, where he teaches courses in composition and rhetoric, technical communication, and professional writing. His book *Writing Workplace Cultures* (Southern Illinois, 2000) was awarded the Distinguished Publication Award for 2001 by the Association of Business Communication. Currently he is composing a hypertext analyzing workplace performance appraisals.

Steven B. Katz is the Roy Pearce Professor of Professional Communication at Clemson University. His publications include *The Epistemic Music of Rhetoric* (Southern Illinois, 1996) and *Writing in the Sciences* (Longman, 2004). His award-winning essay "The Ethic of Expediency: Classical Rhetoric, Technology, and the Holocaust" was reprinted in *Central Works in Technical Communication* (Oxford, 2004).

Bernadette Longo is Associate Professor of Rhetoric at the University of Minnesota. She is the author of the extended cultural study *Spurious Coin: A History of Science, Management, and Technical Writing* (SUNY, 2000). Her current research deals with robots, brains, metaphor, and computer history. Before her academic career, Longo was a contract technical writer for nearly twenty years in the medical and poultry processing industries.

Myra G. Moses received her M.S. in Technical Communication from North Carolina State University. She is currently a Lecturer in the Professional Writing Program in the English Department at North Carolina State University, and is Special Projects Associate for the Service-Learning Program.

Alan Nadel is Professor of Literature and Film at Rensselaer Polytechnic Institute. He is the author of several books on American literature, film, and culture, including *Invisible Criticism: Ralph Ellison and the American Canon* (Iowa, 1988), *Containment Culture: American Narratives, Postmodernism, and the Atomic Age* (Duke, 1995), and *Flatlining on the Field of Dreams: Cultural Narratives in the Films of President Reagan's America* (Rutgers, 1997). Nadel was the recipient of the best essay prize from *Modern Fiction Studies* in 1988 and the best essay prize from *PMLA* in 1993.

Michael J. Salvo is Assistant Professor of Professional and Technical Writing in the Rhetoric and Composition program at Purdue University, where he teaches undergraduate and graduate classes in rhetorical theory, professional writing and technoculture. Salvo's articles have appeared in the *Journal of Business and Technical Communication, Technical Communication*

Quarterly, Computers and Composition, and *Kairos,* for which he was managing editor during the journal's first two years of publication. Contact him at: salvo@purdue.edu.

Beverly Sauer is Professor of Business Communication and Director of the Center for Corporate Crisis Communication in the Graduate Division of Business and Management at Johns Hopkins University, where she teaches risk communication, managerial communication, and cross-cultural communication. Author of the award-winning *The Rhetoric of Risk: Technical Documentation in Hazardous Environments* (Erlbaum, 2003), Sauer has received four National Science Foundation grants to study gesture and workplace communication in the United States, Great Britain, and South Africa.

J. Blake Scott is Associate Professor of English at the University of Central Florida, we he teaches courses in technical communication, rhetoric and composition, and the rhetorics of science and technology. He is the author of *Risky Rhetoric: AIDS and the Cultural Practices of HIV Testing* (Southern Illinois, 2003) and, with Melody Bowdon, of *Service-Learning in Technical and Professional Communication* (Longman, 2003). Scott and Bernadette Longo guest edited the Winter 2006 special issue of *TCQ* on Cultural Studies & Technical Communication.

Jennifer Daryl Slack is Professor of Communication and Cultural Studies in the Department of Humanities at Michigan Technological University. She works in the areas of cultural studies, communication theory, culture and technology, and culture and environment. Slack is the editor of *Animations of Deleuze and Guattari* (Peter Lang, 2003), co-editor of *Thinking Geometrically: RE-Visioning Space for a Multimodal World* (Peter Lang, 2002), and co-editor of *The Ideology of the Information Age* (Greenwood, 1987).

Katherine V. Wills received her doctorate in composition and rhetoric from the University of Louisville in 2004. She teaches composition, technorhetoric, business, technical, and creative writing at Indiana University Purdue University-Columbus. Wills is the Breaking News Editor of *Workplace: A Journal for Academic Labor*. Her work has appeared in *Kairos, Comp Tales: The Composition Frontline*, *Politics of Information: Electronic Mediation of Social Change E-Book* (Eds. Bousquet and Wills, 2004) and *Workplace.*

Index

Action: civic, 2, 108; communicative, 157; ethical, 2, 9, 236, 247; ethical civic, 196; instrumental, 79*tab*, 98*tab*; made possible, 236; participatory methods, 6; possibilities for, 7; postmodern, 236; rhetorical, 224; sanctioning, 225; social, 6, 242; strategic, 79*tab*, 98*tab*
Adams, Katherine, 8
Adler, Patricia, 161
Adler, Peter, 161
Adorno, Theodor, 41, 43, 236
Althusser, Louis, 3, 222, 260; neo-Marxist notion of ideology and, 5; structuralism and, 37
Anderson, Paul, 65
Angell, David, 94
Aronowitz, Stanley, 122
ARPANET, 80–81, 83, 92
Articulation: characteristics of, 38; cultural studies notions of, 11; defining, 38; influence of structuralism on, 37; of meaning, 27; model of communication, 37–42; postmodernism and, 37; process of, 7, 37; of relations of power, 27; relative power of, 38; resistance to rearticulation by, 38; in technical communication, 27; tenacity of, 38; of theories/practices shaping writing, 47; theory, 7
Authors and authorship: authority of, 25, 41; consequences of extending to technical communicators, 42; function of, 25, 40; granting privilege of, 25; in transmission view of communication, 40, 41, 42

Bacon, Francis, 117, 122
Baker, Tom, 138, 142, 147
Bakhtin, Mikhail, 33, 42, 86; chronotopes of, 139, 140; social praxeology of, 139
Barnum, Carol, 50, 64
Baron, Naomi, 97
Barry, Ann Marie, 220
Barry, Dean, 143
Barthes, Roland, 260
Barton, Ben, 4, 13, 119, 120
Barton, Ellen, 153
Barton, Marthalee, 4, 13, 119, 120
Baudrillard, Jean, 118
Baugh, L., 92
Bauman, Zygmunt, 118, 203, 204, 205, 235, 237
Bazerman, Charles, 137, 173
Beck, Ulrich, 172
Bell, Arthur, 120
Benjamin, Susan, 91, 93, 94, 97
Benjamin, Walter, 15, 22, 64, 117, 125, 126, 127

Bennett, Tony, 222
Berger, Peter, 136, 147
Berkenkotter, C., 173
Berlin, James, 6, 74, 196, 201, 219, 220, 225, 229, 234, 235, 246, 272
Best, Jo, 97
Bhatia, Sabeer, 83
Bickford, Donna, 245, 247, 248
Birmingham school, 3, 4, 5, 125, 155, 247
Bly, Robert, 93, 97
Blyler, Nancy, 6, 14, 74, 75, 77, 152, 153
Blythe, Stuart, 6, 164
Boiarsky, Carolyn, 120
Boisjoly, Roger, 114
Bolter, Jay, 75
Boo, Katherine, 215
Boshier, Roger, 72
Boundaries: company specific, 118; organizational, 133; work/play, 76–8
Bourdieu, Pierre, 139, 140, 197, 260, 261; notion of field, 140
Bowdon, Melody, 243
Bradford, Don, 84
Britt, Elizabeth, 107, 108, 120, 133–148, 197
Brown, R.H., 171
Burnett, Rebecca, 120
Bushnell, Jack, 11
Butenhoff, Carla, 44
Butler, Brian, 196, 225, 230–233

Capital: academic, 260; cultural, 7, 262; material, 7; noneconomic, 262
Capitalism: critiques of politics and ethics of, 171; fast, 195, 204; new, 118, 203, 204; postindustrial, 159; technological, 84
Carey, James, 222, 242
Center for Online Addiction, 91
Challenger disaster, 114, 176, 212
Change: agents, 196, 220; creating possibilities for, 151; ethical, 196; methodological, 260; political, 196, 220; social, 153, 158, 196, 220; structural, 173; technical, 220
Charney, Davida, 153
Cintron, Ralph, 156, 158
Clarke, L., 173
Class: cultural studies and, 219; and meaning of culture, 3; positions, 158; relations, 118, 203; rhetorics of, 157; struggles, 4; working, 157
Classification: academic forms of, 261; primitive forms of, 261
Coal Mine Roof Rater system (CMRR), 183, 183*fig*
Collins, Billy, 271, 273
Collins, H., 187
Commercialism, 22
Communication. *See also* Technical communication: articulation view of, 27, 37–42; control over, 220; defining, 26; dependence, 90; distribution of power in, 47; email and, 71–99; "enslavement," 90; ethics and, 62; in ideology of technology, 88–89; interpersonal, 242; mass, 3; models of, 26; neo-Marxist critiques of, 3; as ongoing struggle for power, 34; organizational, 133, 173; origins of in transportation, 28; overload, 98; participatory, 26; politics and, 62; practices, 32, 36, 164; process, 26; professional, 6, 7, 75, 119; redemptive power of, 73; research and, 161, 163–167; social construction of, 71; technical production of, 96; technology, 28, 71; theory, 26–27; translation view of, 26, 32–36; transmission view of, 27–31; as transportation of messages, 28; user-centered methods of development in, 47; verbal/nonverbal, 76
Composition. *See* Writing
Connors, Robert, 9, 29, 30, 35
Constructionism, 139; social, 107, 114
Cooper, Marilyn, 26, 153, 266
Corporation: lack of ties to workers by, 205; as major arena for public life, 205

Couture, Barbara, 267
Creighton, James, 122
Cross, Geoffrey, 138
Crowley, Sharon, 10, 202, 213, 214, 215
Cultural: activities, 7; capital, 7; circuit, 7, 12; construction, 156, 157; contexts, 107; criticism, 115; critique, 196; distinctiveness, 156; forms, 11, 247; geography, 6; institutions, 157; interrelations, 222; "making," 156; objects, 15, 22, 64, 117, 125, 126; power, 23, 24, 117; practice, 5, 156, 157; production, 158, 260; struggles, 21; theory, 108, 157
Cultural studies: articulation theory in, 7; assessment of subject-related effects of power in, 14; as "bricolage," 5; British, 3; class and, 219; coalitions of scholars in, 222; commitment of, 221; concern for knowledge legitimation, 15, 22, 64, 117, 125, 126; contributions of, 11–17; contributions to service-learning pedagogy, 196; cultural circuit, 197; culturalist paradigm, 155; curricula, 264–265; defining, 155, 248; delimiting object of, 124–127; development of approaches to research and curricula, 220; diversity of, 155; driving forces in, 2; emergence/transformation of discourse and, 171; emphasis on discourse, 4; and enhancement of technical communication studies, 1–17; enhancements to, 246–249; extreme usability and, 48; gender and, 219; goals of, 3, 12, 221; imperialist tendency of, 2; insistence of formation of students as critical discursive consumers and, 2; intercultural inquiry, 197; knowledge legitimation and, 112–127; materiality and, 4; as means of reading, 155; models, 197; notion of articulation in, 11; objectivity and, 120–123; overreliance on semiotic analysis, 247; policy participation and, 222; potential pitfalls of, 16; production and consumption of cultural artifacts in, 155; race and, 219; as radically contextual, 2; resistance to hyperpragmatism, 197; rhetorics of culture and, 154–159; service-learning and, 241–256; structuralist paradigm, 4, 5, 155; syllabi, 264; in technical communication methodology, 219–238; technical writing and, 118–120, 151–167, 259–269; through discourse analysis, 116; transactional rhetoric and, 234–238; translation of critique into ethical civic action in, 15; validity and, 120–123; values held by, 222; venues and service-learning in, 265–266
Culture: as analytical focus, 156; artifacts of, 222; as collaboration, 115, 116; complexity of, 62; conscious production of, 259; conservative view, 115; construction of, 222; consumer, 22, 50, 53, 54; corporate, 14, 62; creation of, 156, 157; defining, 5; dominant, 34, 223; expansion beyond one organization, 118; experiential living and, 3; global corporate, 2; hegemonic, 116; historicizing concept of, 3; ideological dimension of, 5; limited views of, 114; "making," 157; material, 155; meaning of, 3; modes of interpretation and, 3; as narrative, 157; nature of, 157; neo-Marxist critiques of, 3; organizational, 115, 196; postindustrial, 225; of professionalism, 201; reproducing, 259; rhetorics of, 151, 154–159, 157; traditions for study of, 3; transformation of, 15; usability and, 61–64; work, 205; working-class, 157; workplace, 6, 200–216

Culture, *Cont'd*
Curran, D.J., 174
Curricula: design change, 200; developments in, 201; formalist, 9; innovations, 201; use of writer's subjectivities in, 212
Cyphert, Dale, 227

Dance, Frank, 73, 88
Darwin, Charles, 260
Dautermann, Jennie, 114, 133
Dawes, R.M., 183
Deans, Thomas, 241, 242
Debs, Mary Beth, 205
de Certeau, Michel, 15, 22, 64, 117, 125, 126, 127, 160, 230
Decoding: dominant, 33; meaning, 28, 32, 33, 34; negotiated, 33; oppositional, 33; in symmetry with encoding, 34; theory of, 32, 33, 34
Deconstruction: as version of structuralism, 5
Deleuze, Gilles, 37, 260
Department of Defense: email development by, 80–81
Derrida, Jacques, 260
Devitt, Amy, 114
Dewey, John, 242
Dilger, Bradley, 22, 23, 47–66
Discourse: analysis, 116, 155; articulated, 116; authorial, 27; authoritative, 234; authors and, 25; communities, 118, 140, 203, 241, 242; creating difference in, 15, 22, 64, 117, 125, 126; cultural studies emphasis on, 4; design, 233–234; dominant, 213; effects on writers, 210–211; genealogy of, 120; historical aspects, 15, 22, 64, 117, 125, 126; humanistic, 41; incitement to, 122; institutional role in production of, 4, 13, 112, 119; knowledge creation and, 171; legitimated knowledge and, 116; mapping, 221; medical, 134; narrating power and authority, 229; as object of inquiry, 124–125; and postmodern inquiry into subjectivity, 203; practices of, 119; professional, 75; racist, 210; revision, 213, 214; role in social relations, 133; scientific, 41, 117, 122; sexist, 209, 210; shared goals in, 220; in social practice, 37, 75; structure in dominance, 34; as struggle mediated by culture, 116; technical, 40, 151–167; theory of, 116; workplace, 187; of workplace production, 223
Doak, Jeffrey, 7, 21, 22, 25–44, 229
Dobrian, Joseph, 89
Dobrin, Sid, 266
Documentation, 231, 232; failure to assess local knowledge, 174; feminist analysis and, 187; limits of, 173; managerial control and, 201; postpromulgation, 175; risk-avoidance, 231; technical, 180–183, 185; web-based, 172; workplace, 201; written, 173
Doheny-Farina, Stephen, 113, 114, 120, 133
Dombrowski, Paul, 173
Domination: economic, 28; imposed/won, 4; political, 28; social, 4
Donoghue, Karen, 49, 53, 54, 55, 56, 57, 59, 66*n2*
Douglas, Mary, 133, 137, 142
Doyle, Aaron, 143
Dragga, Sam, 120, 264, 265
Dubinsky, James, 243, 246
Dubrovsky, V., 72
Dumas, Joseph, 58, 63, 66*n1*
Dunlap, Albert, 205
Durkheim, Emile, 261
Dvorak, John, 51, 55

Ease. *See also* Usability, extreme: cultural power of, 50; defining, 67*n5*; gender and, 51; historicizing, 50–52; ideological power of, 51; "life of," 51; minimization of complexity of prod-

ucts and, 51; role in product design and marketing, 50, 51
Economic: domination, 28; expediency, 13
Efficiency: belief in values of, 82; capitalistic value of, 74; technological imperatives of, 23
Email, 23; ability to work from various locations with, 72; access to, 74; addiction to, 90; blurring of public/private space by, 23, 75–80, 99*n2*; boundaries of work and play and, 71, 72, 75–80; change in communication from, 89, 90; as commercial product, 75; as communication technology, 71–99; constant checking of, 90; conventions, 91–93, 99*n6*; corporate takeover by, 81–83; cultural critique of, 71–99; as cultural practice, 74; democratizing pressure of, 73; dependency on, 91, 97; from desire to increase efficiency, 81–83; effect of ideology on generic conventions of, 73; elimination of social inequities in, 72, 73, 99*n7I*; emoticons and, 76; failure causes, 93; handling large amounts of, 85; hierarchy and, 74; ideology of, 71–99; illusion of control in, 85, 86; impacts on communication patterns and behavior, 75; increases in speed, productivity and efficiency through, 23, 84, 85, 86, 89–96; informality of, 73, 94; instant delivery, 89; interruptions by, 90; junk, 85; lack of formal syntax in, 73; military beginnings of, 80–81; move from traditional institutional framework in, 93–96; organizational limits on, 97; personal, 23, 75, 83–86; personal relations and, 93–96; personal/technological characteristics of, 89–96; in purposive-rational system, 96; rapid expansion of, 91; representation of "freedom" and, 72; reshaping of personal relations by, 83–86; restructuring of conventions in, 93–96; as secure place to participate in discussion, 72; spam, 97; standardized memo form of, 73, 91, 92; styles, 93–96; system identifiers, 91; technical rules of, 91–93; technological relations and, 83–86; time factors, 90
Ericson, Richard, 143, 145, 146
Ethic(s): attention to, 35; of capitalism, 43; commercial, 23; communication and, 62; of ease, 22; of expedience, 6, 9, 23; machine-based, 23; technological, 23
Ethnographies, 200
Ewald, François, 142, 143, 144, 147
Expedience: economic, 13; ethic of, 6, 9, 23; as means and end, 9
Experience(s): embodied, 174; extreme usability and, 9; in highly structured institutions, 184; lived, 203; natural philosophy based on, 122; rhetorical transformation of, 184–185; shaped by language practices, 203; subjective, 173; as warrant for risk decisions, 173, 174

Faber, Brenton, 196, 225–227
Faigley, Lester, 4, 13, 119, 203
Farrell, Nick, 97
Federal Mine Safety and Health Act of 1977, Public Law 91-173, 176–189, 189*n4*
Feenberg, Andrew, 220
Festa, Paul, 72
Finholt, Tom, 72
Fish, Stanley, 137
Fiske, John, 28
Fitts, Karen, 6
Flower, Linda, 250
Foley, Douglas, 156, 159
Foucault, Michel, 3, 4, 6, 7, 21, 40, 41, 86, 107, 125–126, 260; archeological method of, 4, 13, 119, 124; on authorship, 25; and Birmingham school, 4, 5; critical theory of, 4, 13, 119; on discourse, 116; on exercise of power, 4, 13, 119; notion of geneal-

Foucault, Michel, *Cont'd*
ogy of, 15; object of study as difference and discontinuity and, 127; poststructuralist theory of knowledge/power, 4, 247; resistance to repressive power, 13; on subjugated ways of knowing, 15, 22, 64, 117, 125, 126; on systematic history of discourses, 120, 125, 126
France, Alan, 6
Frankfurt school, 3
Frederick, Christine, 50, 51, 54, 63
Freed, Richard, 118, 119, 120
Freire, Paolo, 242
Freud, Sigmund, 260

Garnham, Nicholas, 222
Gee, James, 204
Geertz, Clifford, 187
Geise, Mark, 81, 99*n3*
Geisler, Cheryl, 76
Gelernter, David, 51
General Accounting Office (GAO), 115
Genres, 139, 140, 141, 143, 173
George, Diana, 271–274
Gere, Ann, 245
Giroux, Henry, 14
Globalization, 118, 200, 203, 204, 205, 227
Gordon, Colin, 143
Gordon, D., 122
Goswami, Dixie, 204
Gould, John, 66*n1*
Grabill, Jeffrey, 2, 6, 17*n1,* 108, 151–167, 272, 273
Graham, John, 89
Gramsci, Antonio, 3, 15; concept of articulation and, 37; on hegemony, 4; notion of organic intellectual, 4; notions of hegemony, 125
Griffeth, Roger, 120
Grossberg, Lawrence, 2, 5, 13, 14, 17, 22, 36, 37, 40, 121, 155, 221, 222, 242, 247, 248
Guattari, Félix, 37

Habermas, Jürgen, 71, 72, 73, 77, 78, 86, 97, 157; on acceptance of scientific/technological advances, 76; Chart of Work and Interaction, 79, 79Itab; conditional imperative of, 90; on "conditional predictions," 89; critical theory of, 96; on modernism, 236; purposive-rational ideology of, 96, 97, 98*tab*; rationality and, 237; theory of communication/work, 74, 75, 79, 83
Hafner, Katie, 76, 81, 92
Hall, Stuart, 3, 4, 5, 11, 22, 32, 34, 37, 125, 154, 222, 246
Hammersley, M., 161
Hardy, Ian, 81
Harvey, David, 118, 203, 266
Hayano, David, 209, 211
Hegemony, 125; Gramsci on, 4; as shifting power struggle, 4
Henry, Jim, 2, 13, 14, 195, 196, 199–216, 225, 227–228, 246
Henson, Leigh, 241, 243, 246
Herndl, Carl, 2, 6, 11, 14, 15, 17, 17*n1,* 73, 96, 114, 120, 152, 153, 154, 160, 173, 206, 213, 220
Herzberg, Bruce, 119, 245
Heteroglossia, 33, 42
Hocking, John, 122
Holdstein, Deborah, 74
Hoover, H.C., 184
Hoover, L.H., 184
Horkheimer, Max, 41, 43
Hotmail, 83, 84, 85, 86, 87
Howe, Peter, 77
Huckin, Thomas, 241, 242, 243, 246
Huff, Chuck, 72
Hull, Glynda, 204
Human Genome project, 260
Humanism, 8
Huntley, D.W., 183
Hurston, Zora Neale, 207
Hyperpragmatism, 7–11, 196, 242; conservatism of, 10; co-optation of service-learning potential by, 243, 244,

245; defining, 9; devaluation of ethical intervention in, 14; emergence of, 9; goals of, 9, 13; liberal approaches, 10; limitations of, 11–17; regulatory power and, 12; resistance to, 197; rhetorical/social forms of, 10; service-learning and, 242

IBM, 81, 82, 83
Identity: autoethnographic inquiry into, 206–213; cultural agreement on, 39; cultural struggle over, 22; defining, 37, 38, 39; forging, 37; organizational, 113; in social formation, 22; work, 206–213
Ideology: addiction as, 90; and development, 86–89; of ease, 48; of email, 71–99, 73; means-end, 91; modernist, 211; neo-Marxist, 5; products of, 86–89; of technology, 75, 76
Imaginaries: insurantial, 142, 143, 148
Inkster, Robert, 92
Institutions: in analyses of discursive practices, 133; bureaucratic, 173; capitalist, 171; compared with organizations, 134, 135; constraints of, 3, 12; as cultural agents, 112, 113, 147, 157; cultural critique and, 138; cycle of documentation in, 185; emergence of, 136; evolution of frameworks of, 78, 79*fig*, 79*tab*; influence on discourse, 112; insurance as, 141–147; interrogating, 147; justifying principles of, 136, 137; mapping, 221; meaning of, 134; nonproductive, 157; organization of operations of, 111; practices of, 108; reification of, 136; reliance on rhetoric to objectify existence, 137; as rhetoric, 147–148; rhetorical work of, 6, 133–148, 151, 159–167; role in production of discourse and knowledge, 4, 13, 119; as scene of struggle, 159; social origins of, 137; space of agency and, 159–167; taken for granted, 134–137; technological, 47; technology absorbed into framework of, 78; visibility of, 47
Insurance: complexity of, 142; defining, 142; as form of governance, 143; legitimacy of, 142; material consequences of, 143; as moral technology, 145, 146; political economy of, 146; risk and, 142; technical writing and, 143–147; as technology, 142
Intertextuality, 115
Irving, Blake, 85
Irwin, A., 172

Jacobson, Robert, 220
Jaffee, Dwight, 216
Jameson, Fredric, 3, 12, 118, 203, 236, 237
Johnson, John Butler, 111
Johnson, Richard, 7, 12, 14, 247
Johnson, Robert, 14, 47, 55, 57, 64, 220, 223
Johnson-Eilola, Johndan, 203, 222, 267

Kanfer, Alaina, 72, 97, 99
Katz, Steven, 6, 9, 13, 17, 23, 43, 44, 71–99, 120
Kaufer, David, 196, 225, 230–233
Kells, Michele, 265
Kleimann, Susan, 115
Knorr, Karin, 172
Knowledge: conceptual, 56; creation, 171; cultural power of, 112; delegitimated, 117; dominant, 125; extradiscursive operations of, 4; institutional role in production of, 4, 13, 119; instrumental, 56; legitimation of, 6, 13, 107, 111, 112–118, 120, 133, 171; limiting, 107; local, 172, 174, 178, 183; making, 24, 137, 260; management, 220; marginalized, 4, 13, 111, 112–118, 119; naive, 118; nature of, 214; organization through science, 117; possession of, 7; pragmatic, 59; production, 7, 166; qualitative, 111; rationalized, 22, 23;

Knowledge, *Cont'd*
reconceptualization of, 118; scientific, 8, 23, 117, 122, 140, 178; silencing, 116; situated, 119; specialized, 58; of struggles, 117; subjugated, 4, 15, 22, 64, 117, 125, 126; subordinated, 125; systems, 120; technical, 7, 22, 43; theory of, 4; transmission of, 28; useful, 7; Victorian notions of, 260; women's, 187
Koku, Emmanuel, 72
Kossek, Ellen, 120
Kynell-Hunt, Teresa, 8, 9, 196, 225, 228–230

Labor: alienated, 157; communicative, 157, 158, 160; flexible, 118, 195, 200, 203, 205; off-shore sources, 176; policy, 223; sensuous, 157
Laclau, Ernesto: concept of articulation and, 37
Lamb, Linda, 76, 91
Language: construction/mediation of realities through, 214; context free, 97; of encoding/decoding, 37; free of traditional social content, 94; functional, 230; local, 226; natural, 49; philosophical, 230; of practice, 117; production, 22; reality formation and, 203; shared, 78; of translator, 37
Lankshear, Colin, 204
Lather, Patti, 122
Latour, Bruno, 171, 172, 173, 234, 259
Lawson, Hilary, 111
Lay, Mary, 120
Lee, Judith, 71, 76, 93, 94, 95, 99*n71*
Leitch, Vincent, 3, 12, 112, 137
Levelt, W.J.M., 184
Levi-Strauss, Claude, 5
Lewis, Clayton, 66*n1*
Lindquist, Julie, 157, 158
Linnaeus, Carl, 260
Lionnet, Françoise, 207
Logic: Aristotelian, 122
Longaker, Mark, 17*n1*
Longo, Bernadette, 1–17, 4, 6, 7, 8, 9, 11, 13, 47, 61, 62, 64, 107, 111–127, 133, 140, 156, 171, 220, 259
Louis, Meryl, 136
Luckmann, Thomas, 136
Lynch, M., 172
Lyotard, Jean-François, 116, 125; notion of grand narratives, 118; *petit recits* of, 118, 203, 211, 212; theory of discourse, 116

Making: acts of, 158; arts of, 167*n6*; culture, 156, 157; knowledge, 260; meaning, 3, 4, 7, 34, 36, 222; sense, 156
Management, scientific, 201
Mark, C., 184
Markel, Michael, 30
Marx, Karl, 157
Marx, Leo, 71
Marxism, 3, 4, 5, 89, 121
Materiality: cultural studies emphasis on, 4; of power, 5
Matthews, Catherine, 243
Maxwell, Lee, 49
Mayhew, Deborah, 49, 58, 59, 61, 62
McAlpine, Rachel, 64
McAteer, J.D., 177
McBeth, Sally, 200
McCarthy, Lucille, 120
McCloskey, D., 183
McCormick, N.B., 72
McQuail, Denis, 28, 222, 225
McRobbie, Angela, 221
Meaning: articulation in practice, 37; articulation/rearticulation, 21, 27; changing conceptions of, 27–42; circuit of, 37; constitution of, 27, 32; construction of, 32; conveyed, 21; cultural criticism and, 115; dominant, 34; elusiveness of, 34; empowerment of, 22; encoding/decoding, 28, 32, 33, 34; as fixed entity, 29; fluidity of, 36; in

intention of sender, 31; interpretation of, 21; location of, 30, 34; making, 3, 4, 7, 34, 36, 222; power in making, 34; power to fix, 40; preferred, 34; produced through interaction of sender and receiver, 32; production of, 11, 22; in social practice, 37; transfer of, 31; transmission of, 28
Mediators, 37
Microsoft, 81, 83, 84, 86, 87
Miles, Libby, 6
Miller, Carolyn, 8, 43, 83, 96
Miller, David, 7, 21, 22, 25–44
Miller, Susan, 203, 225, 246
Mine safety, 171–189
Mirel, Barbara, 48, 49, 56, 57, 65
Modernism: resistance to totalizing narratives of, 203
Molinda, G., 183
Morgan, Gareth, 135
Morley, David, 33
Moses, Myra, 23, 71–99
Multiculturalism, 120

Nahrwold, Cynthia, 2, 152, 153, 154, 160
Narratives: culture as, 157; of globalization, 227; legitimating, 136; master, 173; of modernism, 203; resistance to master, 211
National and Community Service Trust Act of 1993, 241
Nelson, Cary, 5, 17, 121, 155
Nicholson, Joel, 120
Nielsen, Jakob, 48, 49, 52, 53, 54, 56, 58, 62, 65, 66*n3*
Noah, Lars, 175
Norman, Donald, 51, 56

Object: in cultural context, 125; as discourse, 124–125; in historical context, 125–126; as ordered, 126
Objectivity, 120–123; defining, 121; in research, 153
Odell, Lee, 204
Olson, D.R., 173, 187
Organizations: class status in, 207–208; compared with institutions, 134, 135; cultural contexts and, 107, 115, 136; defining, 135; dynamics of, 107; effect of values on document review practices, 115; expectations of, 244; formal, 135, 148*n1*; functioning of, 135; goals, 135; improving effectiveness of correspondence in, 137; insurance, 134–148; operation of, 44; politics of, 44; social, 135; stability of borders, 135; structures of authority in, 135; writer status in, 207–208
Orlikowski, Wanda, 76, 95

Paradis, James, 133
Pedagogy, 195–197; action-oriented, 225; antirhetorical, 8; autoethnographic inquiry in, 195, 206–213; critical, 17*n1,* 196; current-traditional, 202; efficient models of, 10; ethnographies in, 200; hyperpragmatist, 9; implications for, 42–44; liberatory, 16, 17*n1*; to position students to enact ethical change, 196; process, 195, 202, 203; radical, 6, 242; service-learning, 196, 197, 212; strategies for combining cultural studies and technical communication, 263–268; technical writing, 259
Phronesis: in service-learning, 242
Pluralism, 34
Political: change, 73, 196; commitment, 2; domination, 28
Polysemy, 33, 34, 35, 36
Pomerenke, Paula, 138
Porter, James, 2, 6, 13, 14, 114, 120, 135, 137, 148, 148*n2,* 153, 159, 236, 247, 248
Positivism, 12; cultural emphasis on, 173; scientific, 107, 171, 173
Postmodernism, 120; in classroom content, 267, 268; composition and, 202–203; impact on research, 152; influence on articulation, 37; literacy

Postmodernism, *Cont'd*
and, 237; references to textual/social practices, 203; relations with authoritative discourse, 234; terminology of, 203
Poststructuralism, 107, 121, 247; of Foucault, 4
Power: academic, 261; of academic capital, 260; authorial, 22; capitalist notions of, 172; changing conceptions of, 27–42; communication as struggle for, 34; critiquing structures of, 247; cultural, 24, 117; differential relations of, 34, 35; disabling, 13; disciplinary, 4; discursive, 5; displacement of, 34; distribution of, 47; ethical exercise of, 253; extradiscursive operations of, 4; to fix meaning, 40; flows, 221; hegemonic forms of, 6, 197; infiltration of, 13; institutional, 4, 13, 108, 119, 266; institutional role in production of, 4, 13, 119; invisibility of relations of, 63; in making meaning, 34; materiality of, 5; negotiated, 32; operations in organizations, 140; regulatory, 12; relations of, 1, 27, 108, 109, 140, 141; of senders of messages, 28; sources of, 11; structures, 73; subject-related effects of, 14; successful communication and, 29; systems, 23, 120; texts as sources of, 5; theory of, 4, 22; through design of space, 6
Practice(s): access, 161; communication and, 161, 163–167; culturally specific origins of, 171; historically contingent, 4; initiation, 161; local politics and, 161; participation, 161; sustainability and, 161; tactical nature of, 160
Pratt, Laurie, 94
Pratt, Mary Louise, 208
Production: commercial, 23; consumer, 5, 22, 50; ideology and, 86–87; knowledge, 7; language, 22; material, 3; of meaning, 11, 22, 33; symbolic, 3
Productivity: belief in values of, 82; capitalistic value of, 74, 77; economies of scale and, 201; email and, 23, 99; increasing, 76; scientific management and, 201; subsystems of, 78; technological imperatives of, 23; textual studies of, 158; usability and, 51

Quisenbery, Whitney, 48, 49, 65

Rationality, 237; critiques of, 171
Reed-Danahay, Deborah, 213
Reich, Robert, 204, 205, 266
Relationships: affecting knowledge legitimation, 6, 7; author/discourse, 25; building, 164, 168*n8*; cultural theories and, 24; institutional, 119, 120; intercompany, 118; knowledge/power, 112; meaning/power, 37; means-end model, 23; of object and one who studies, 127; people/products, 23; personal, 23, 161; personal/business, 24; researcher-participant, 6; spatial, 184; text/context, 133; work, 23
Research, 107–109; community-based, 160; critical turn in, 153; criticism and, 151; cultural/rhetorical intersections in, 166; cultural studies and, 5; descriptive, 2, 152; and development, 86–89; distinction between "method" and "methodology," 153, 154; ethical action and, 2; in formal organizations, 135; goals of, 153; for health and safety of workers, 171–189; impact of rhetoric and postmodernism on, 152; instrumentalist, 2, 152; insurance organizations and, 137–148; intervention and, 2, 152; location and, 159, 160, 166, 167; methodological issues, 152–154; objectivity in, 153; organization analysis in, 137; organizing activities

of, 160; political stance in, 153; practice, 161–167; production of culture and, 167; rhetorical agency and, 151–167; social constructionist, 3, 6, 12; taking sides in, 153; technical writing and, 151–167; theoretical imperialism and, 2; usability testing, 57–64
Reynolds, Nedra, 245, 247, 248, 272
Rhetoric: analytical power of, 158; of class, 157; classical, 6; of culture, 157; encomium, 233; impact on research, 152; institutional practices and, 108; of institutions, 147–148, 159–167; judicial, 233; legislative, 174; policy, 233; productive, 225; as productive art, 167*n6*; as productive technology, 219–238; traditional, 8, 10; transactional, 229, 234–238
Rhodes, John, 62
Rickard, T.A., 8
Risk, 142, 144, 145; assessment, 173, 174, 180; decision-making, 172, 173, 180–183; distribution, 178; formal discourse of, 183; management, 173; normalization of, 173; prediction, 176; public understanding of, 172; regulation, 174, 178
Rorty, Richard, 121
Rouse, Joseph, 188
Russell, David, 201

Salvo, Michael, 2, 14, 196, 219–238, 248
Sammons, Martha, 64
Sauer, Beverly, 16, 108, 109, 120, 171–189, 196, 225, 230–233
Savage, Gerald, 196, 225, 228–230
Schein, Edgar, 216
Schryer, Catherine, 108, 138–148, 197
Schultz, Aaron, 245
Schultz, Heidi, 90
Schwartz, John, 223
Science: formation of legitimated language of practice and, 117; knowledge organization through, 117; method-driven, 122; neutrality of, 43; political/economic dimensions of, 172–175; positivist view of, 8; postpositivist, 171; public, 122; sociocultural studies of, 172–175; technical writing and, 8; theory of, 173; what counts as, 172
Scott, J. Blake, 1–17, 4, 7, 8, 166, 196, 197, 241–256, 266
Segal, Judy, 134
Selfe, Cynthia, 152, 203, 212
Selfe, Richard, 152, 203, 212
Self-reflexivity, 2, 13, 35, 122, 124
Selwyn, Neil, 72
Selzer, Jack, 115
Semiotics, 5
Service-learning, 241–256; academic skills and, 243; action plan in, 253–256; benefits of, 242; contextualized learning in, 241; critiquing structures of power and, 247; cultural studies enhancements and, 246–249; defining, 241; discourse analysis and, 252, 253; ethical action and, 247; final report in, 253; hyperpragmatism and, 242; interpersonal communication and, 242; models, 245; organizational politics and, 242; pedagogical applications of, 249–256; phronesis and, 242; potential of, 241–243; practical demands of, 244; as "practice," 244; pragmatism and, 242; projects, 244–246, 249–256; proposal assignment, 250, 251, 252; reflection and, 243, 245; unfulfilled promise of, 243–246
Shannon, C., 28
Silverman, David, 135, 148*n1*
Simplicity: extreme usability and, 53
Simulacra, 118
Sirc, Geoffrey, 237
Slack, Jennifer, 7, 21, 22, 25–44, 229
Smart, Graham, 115
Smith, Jack, 83
Social: action, 6, 242; change, 3, 73, 108, 153, 158, 196, 220;

Social: action, *Cont'd*
constructionism, 10, 107; discourse, 75; domination, 4; hierarchies, 94; interaction, 74; life, 136; norms, 78, 80, 92, 94; organization, 135; praxeology, 139; praxis, 10; process, 10, 96, 200; reform, 242; relations, 133; responsibility, 22, 24; system, 78; theory, 108; values, 3
Souther, James, 30, 35
Spilka, Rachel, 10, 134
Spooner, Michael, 73, 74, 75, 84, 96
Sproull, Lee, 72
Stacks, Don, 122
Standards, regulatory: accumulation of scientific and local knowledge in, 178; across diverse worksites, 177; collusion in, 174; continual adjustment of, 173; defining, 178, 179; enforceability of, 177; expunging old, 185; in hazardous environments, 176; heterogeneity of, 178; history of risk assessment in, 180; interpretations of intent in, 175, 176; in large regulatory agencies, 178–187; limits on toxic substances, 178; as living documents, 109, 172–189; local knowledge and, 174; ongoing reinterpretation, 179–180; reexamination of after accidents, 177; regulatory framework for, 176; resistance to by social systems, 186–187; revision, 172; role of social science theorists in creating, 173; specification of testing methods in, 178; unscrupulous inspectors and, 174
Stewart, Kathleen, 156, 157
Strate, Lance, 72
Structuralism, 139, 155; Birmingham school and, 4, 5; cultural studies and, 5; influence on articulation, 37; linguistic turn of, 5
Students: cases/real-world assignments in courses, 10; as change agents, 220; as cultural critics, 2; ethical change in workplace and, 196; feelings on activism, 245, 246; motivating, 243; preparation for postmodern workplace, 259; as rhetorically effective producers, 2; of technical writing, 259–269; unbalanced education in technical schools, 29
Stygall, Gail, 115
Subjectivity: postmodern inquiry into, 203; radical, 236; reshaping through technology, 203; of writers, 205
Sullivan, Dale, 6, 15, 17, 43
Sullivan, Patricia, 2, 6, 13, 14, 153, 159, 236, 248
Sutliff, Kristene, 241, 243, 246

Taylor, Frederick, 201, 205
Technical communication: abstraction in, 267; as academic discipline, 36; alternative theoretical framing in, 266–268; articulation model, 21, 22, 37–42; cases/real-world assignments in courses, 10; change in theory and practice of, 40; civic action and, 2; collaboration in, 268; conceptualization of between public and private ownership, 152; concern with formal organizations, 135; corporatization and, 10, 11; creating meaning through power links in, 7; critique and, 10, 11; as cultural activity, 7; cultural studies and, 1–17, 219–238, 259–269; curricula, 264–265; curricula in, 213–214; defining, 223, 224; design of living documents in, 175–178; downplaying work of culture and, 62; effect on future risk decision-making, 180–183; email and, 71–99; emphasis on student development, 245; ethical dimensions of, 6; expansion of focus of, 107; experimentation in, 268; and extreme usability, 47–66; functional practices in, 21; funding and curricular ties to industry and, 10; hyperpragmatism

in, 7–11; ideologies regulating, 3, 12, 75–80; implications for pedagogy; influence of World War II on, 9; institutional power dynamics and, 108; instrumental practices in, 21; mechanical nature of earliest courses in, 8; methodology, 219–238; modern courses in, 9; narrowing of agency by global corporate culture, 2; nature of content of, 214; normative goals of, 1; organizational analysis in research, 137; organizational approaches to, 147; as part of power/knowledge formations, 3, 12; pedagogy, 1; personal email and, 72; as phronesis, 6; policy insights and, 222, 223; political factors in, 6; power dynamics of, 13; practices, 23; pragmatism and, 1, 242; as praxis, 6; professionalization of, 35; for public good, 1; regulatory power and, 12; research methods, 108; research on insurance organizations and, 133–148; rhetoric and, 6; role in hegemonic power relations, 1; service-learning approach, 241–256; social dimension of, 3, 12; sociocultural studies of, 172–175; syllabi, 264; systems thinking in, 267–268; theorizing in, 268; transformations of, 12; translation model, 21, 22, 32–36; transmission model, 21, 22, 27–31; venues and service-learning in, 265–266

Technical communicators: as authors, 22, 23, 25–44, 229; clarity and brevity goals for, 31; disparate images of, 26; education of, 43; ethic of ease and, 23; exercise of power by, 36; expertise issues, 228–230; familiarity with technical field of sender, 35; importance of ethics for, 43; limited view of culture by, 61; mobilization on one's own behalf, 195; need for understanding of language, 35; need to learn technical content of work, 43; as neglected population, 209–213; public life of, 205–206; requirements of, 35; role as mediator between technicians and users, 114; as "support staff," 207–208; as surrogate engineer, 29, 30; as "talking handbooks," 208–209; as translator of meanings, 25, 32–36; as transmitter of messages, 25; transparency of, 30; updating/expunging regulations and standards, 186–187–187

Technical writing. *See also* Technical communication: absence of concept of culture as focus in, 156; affinity for scientism, 259; ambivalence in practices, 166; attributions in, 21; authorship in, 21; changing practices in, 165, 166; clarity of, 111; class position and, 167*n5*; as communicative labor, 158; control of scientific systems by, 111; critical research in, 152–154; cultural contexts of, 112–118; cultural studies and, 111–127, 151–167, 259–269; development of, 112; discursive nature of, 125; discursive process change in, 200; early pedagogy, 8; encoding and, 29, 30, 32, 33, 34; encounter with researcher and, 127; as engineering writing, 29; evolution of, 30; flexibility of labor and, 200; as hegemonic tool, 7; institutionalization as discipline, 10; in insurance organizations, 137–148, 143–147; invisibility of, 111; journalistic style in, 53; legitimation of knowledge through, 6, 112–118; maintenance of class privilege and, 8; as mediator between technology and users, 117; as mundane practice, 125; in organizational charts, 207–208; outdated curricula in, 260; participation in institutional relationships, 117; professional assimilation and, 9; public meetings and, 165; purity of, 8; reports, 165; research, 111–127;

Technical writing, *Cont'd*
role in culture, 113; scannability and, 53; service-learning, 241–256; as situated practice, 2; in the social factory, 151–167; in system of knowledge and power within culture, 118; teaching, 259–269; transmission and, 29; usability testing and, 60, 61; use of writer's subjectivities in curriculum, 212; "value added" issue, 207–208; vocational needs and, 9
Technoculture, 220; demands of, 225; deployment of technology as, 223; production of artifacts of, 220
Technology: absorption into institutional framework, 78; autonomy of, 47; communication, 28, 71; computer, 22, 51; control of over life, 77; demands for ease in, 56; demands on existing standards by, 173; deployment of, 223; digital, 220; distribution of power in, 47; email, 71–99; extreme usability and, 52–57; ideology of, 75, 76; innovation and feature application, 88; innovation and ideological products, 87–88; of justice, 143; mobile, 76; moral, 145, 146; PDA, 77; post-War boom in, 9; productive, 51, 219–238; reshaping subjectivity through, 203; rhetoric as, 219–238; social construction of, 71; subsystems of, 78; transfer, 113; uncritical acceptance of, 56; understanding, 48; user-centered methods of development in, 47
Terminello, Verna, 91, 93
Texts: alternative ways of reading, 155–156; connection to cultural practice, 5; failure to mediate technology to users, 114; in institutional positions, 5; open, 33; reference, 227; relation to contexts, 133; resonancy inscribed in, 115; as sources of power, 5; technical, 11; technical communication, 225–234
Theory: articulation, 7; communication, 26–27; of communication/work, 74; conservative, 10; critical cultural, 108, 152, 157; of culture, 157; of disciplinary power, 4; of discourse, 116; encoding/decoding, 32, 33, 34; feminist, 6; institutional, 137; of knowledge, 4; macroeconomic, 17*n1*; Marxist, 3; of narrativity, 157; neo-Marxist, 17*n1*; organization of practice through, 117; of power, 4, 22; radical education, 6; rhetorical, 158, 227; of science, 173; social, 108; social constructionist, 10; of transmission, 28; of usability, 52–53, 58
Thornton, Alinta, 72, 73, 88
Thralls, Charlotte, 115
Three Mile Island event, 176
Time management, 77
Timmons, Theresa, 138
Tomlinson, Hugh, 121
Tomlinson, Ray, 81
Treichler, Paula, 5, 17, 121, 155
Tschabitscher, Heinz, 90

Ulmer, Gregory, 220
University: corporatization of, 10
Usability: automatic, 57–61, 60; brevity and, 53; condensations of, 49; consumer culture and, 50; costs of, 54, 57; culture and, 61–64; defining, 48–52; demand effect on software, 59, 60; easily remembered interface and, 48; effect of World War II on, 50; efficiency of use and, 48; errors and, 48; extreme, 22; feedback/reporting technologies, 60; focus on commerce, 53; historicizing, 50–52; implementing, 57; increased importance of, 47; integration of, 64; Internet and, 53; learnability and, 48; maturing of, 47; models of, 23; multiple dimensions of, 49; online, 66*n3*; opponents of, 51; oversimplification of, 49; pragmatism of, 52; pressure to

make quantitative, 63; principles to encourage development of, 66*n1*; product, 23; role in electronic design, 47; satisfaction in use and, 48; technical idiocy and, 55, 56; testing research, 57–64; theories of, 52–53, 58; traditional, 22
Usability, extreme: "best practices" approach and, 47, 65, 66; correction of novice shortcomings in, 56; defining, 48–52; facilitation of enjoyment and, 54; focus on ease in, 49; high-tech fields and, 48; historicizing, 50–52; lack of conceptual knowledge in, 56; limitation of usability to expediency and, 47; limitation of visibility of technology and, 47; meaning of extreme and, 52; as methodology, 57–64; minimal training aspect, 49; needs of business shaping user experience in, 55; as quality possessed by technology, 52–57; reduction of user engagement and, 47; rejection of complexity and, 52; as replacement for usability, 47; repression of critical use of technology and, 52; reproduction of ideology of ease and, 48; selectivity of, 63; technical communication and, 47–66; undermining of agency of customers and, 54m55; use of natural language in, 49; user experience and, 54, 55
Utilitarianism, 8

Validity, 120–123; cultural studies and, 120–123; poststructuralist view, 122
Values: social, 3
Van Mannen, John, 161, 216
Vaughan, David, 36
Vaughan, Diane, 173
Virillo, Paul, 203

Wacquant, Loic, 139
Warren, Carol, 161
Watkins, Evan, 54, 97
Weaver, W., 28
Weber, Max, 135
Weblogs, 66
Weil, Debbie, 90, 91
Weisser, Christian, 266
Wells, Susan, 43, 44
Williams, Raymond, 3, 4, 28, 39
Willis, I.J., 184
Willis, Paul, 156, 157
Wills, Katherine, 1–17, 197, 259–269
Wilson, Greg, 14, 197, 266, 267
Windahl, Sven, 28
Winsor, Dorothy, 173
Workplace: changes in, 204–206; discourse communities, 242; discursive networks and, 227–228; flexibility of labor in, 205; ownership of, 205; preparation of students for, 259–269; production discourse, 223; rewriting cultures of, 213–216; union membership and, 205; writing, 241
World Wide Web: availability of email on, 84; optimism surrounding introduction of, 73; relation to democracy, 73; ubiquity of, 225; United States' dominance on, 72, 73
Writing: administration, 213; autoethnographic inquiry, 206–213; composition, 200–204; contributions to products, 201; discursive networks and, 227–228; emergence of term papers in, 201; regulatory, 230–233; rhetorically sensitive, 242–243; shift in emphasis from product to process, 200–204; specialization and, 201; turf battles in universities over, 201; workplace, 199–216, 241
Wurman, Richard, 220

Yates, Joanne, 91, 201

Zak, Michele, 120
Zappen, James, 122
Zimmerman, Beverly, 243
Zuccermaglio, Cristina, 73, 92, 93, 94